"In *Noah's Flood Revisited*, Hugh Ross breaks down one of the most controversial topics in history. As both an astrophysicist and an apologist, he makes a compelling case that the Genesis flood account can be understood as a historical event consistent with both Scripture and science. Ross's rigorous approach provides a valuable framework for honestly engaging this challenging subject. This book is essential reading for anyone seeking to reconcile faith and science regarding Noah's flood."

—Mike Signorelli
Lead Pastor and Founder of V1 Church
Author and YouTuber

"No one bridges the gap between science and Scripture better than Hugh Ross. His new book on Noah's flood will prove helpful to both skeptic and believer."

—Dr. Ben Young
Senior Pastor, Second Baptist Church of Houston
Professor and Author

"Ten years ago, this book could not have been written. The advances in science and understanding of the ancient world have created powerful evidence for the date and extent of Noah's flood. Dr. Hugh Ross, with a deep commitment to biblical authority, brings clarity from both Scripture and science to the controversial issues related to Noah's flood. He builds layer after layer of commonsense, well-researched evidence for a credible account of Noah's flood. This book is a deep contribution to a long conversation about Noah's world."

—Mark Ashton
Lead Minister, Christ Community Church

"The question of the historicity of Noah's flood is often presented as a test case by scientific thinkers who are also critics of the Christian faith. Hugh Ross, with characteristically erudite argumentation, persuasively demonstrates that a close, clear, and faithful reading of the biblical text of the book of Genesis simultaneously holds up to earnest scientific inquiry and the larger truth claims of Christianity. This book is essential reading for anyone who seeks to explore the relationship between the Bible, faith, and scientific inquiry. *Noah's Flood Revisited* offers an extraordinary window into understanding the veracity, scope, nature, and purpose of the biblical account of the flood and what it can teach us about the nature of humanity, our world, our history and future, and the eternal faithfulness of God who is enthroned above it all."

—Corné J. Bekker
D. Litt. et Phil., Dean and Professor
Regent University School of Divinity

"For almost four decades Hugh Ross has brought to the church the orthodox understanding of Scripture as special revelation, and affirmed with general revelation from nature to, as Kepler put it, 'discover the *rational order* and *harmony* which has been imposed on it by God.' God's Word and his world must tell the same story because he is the Author of both. This is necessary for both evangelism and apologetics, to reach our scientific culture with the gospel, and to teach our youth what is real and true in creation. As a geologist and geochemist, I thank the Lord for Hugh Ross's skill at communicating modern geoscience so well. His four-page description of plate tectonics is enough to confirm that Noah's flood certainly was not global. This book is a must-read for every Christian addressing the nature of Noah's flood."

—Ken Wolgemuth
Founder, Solid Rock Lectures
Coauthor, *The Grand Canyon, Monument to Ancient Earth: Can Noah's Flood Explain the Grand Canyon?*

"This is my favorite of Hugh Ross's books on science and faith. As I read through it, I found myself enjoying it more and more. It's so refreshing to read an apologetics book that uses rigorous logic and yet shows a humble respect toward both the biblical text and the supporting modern science. Hugh Ross has a deep understanding of both. And I know of no other evangelical leader who has Ross's ability to explain the sometimes difficult science in such easy-to-understand language.

Many readers will be surprised, as I was, at the breadth of historical, linguistic, and scientific evidence that Ross brings together that supports the biblical account of the flood. But more importantly, he shows that no one need give up biblical inerrancy nor modern science when considering the biblical account. As such, it will be an immensely useful apologetics tool when encountering skeptics who argue that the flood contradicts modern science.

In short, I know of no other scholarly work that provides such a logical and clear explanation of how the biblical account of the flood is consistent with science. This book should prove to be a transformative resource for those who want to provide logical and scientifically accurate reasons for their belief in the Bible's description of the flood."

—Michael E. Summers
Professor of Planetary Science and Astronomy
George Mason University

HUGH ROSS

NOAH'S FLOOD REVISITED

NEW DEPTHS OF INSIGHT FROM SCIENCE AND SCRIPTURE

rtb PRESS

Cover design: Danielle Camorlinga
Interior layout: Christine Talley

Names: Ross, Hugh N., author.
Title: Noah's flood revisited : new depths of insight from science and scripture / Hugh N. Ross.
Description: Includes bibliographical references and index. | Covina, CA: RTB Press, 2025.
Identifiers: ISBN 978-1-956112-08-5
Subjects: LCSH Bible and science. | Bible. Genesis--Criticism, interpretation, etc. | Creationism. | Evolution--Religious aspects--Christianity. | Noah (Biblical figure) | Noah's ark. | Deluge.| BISAC RELIGION / Creation | RELIGION / Religion & Science | RELIGION / Theology | RELIGION / Christian Theology / Apologetics
Classification: LCC BS650 .R67 2025 | DDC 261.5/5--dc23

Printed in the United States of America

First edition

1 2 3 4 5 6 7 8 9 10 / 29 28 27 26 25

For more information about Reasons to Believe, contact (855) REASONS / (855) 732-7667 or visit reasons.org.

Contents

List of Figures and Tables

Figures

Tables

Acknowledgments

Nearly every day, I post briefly on social media about science and how it connects with biblical truth. The posts that stir up more engagement than any other—by far—are those touching on Noah's flood. Their questions, comments, and exchanges helped identify and prioritize the issues addressed in these pages.

Several researchers, scientists, pastors, and students in my Paradoxes class also provided a sounding board for me as I developed content. Their questions, critiques, and suggestions shaped much of the book's content. I owe a great thanks to all.

Further help came via the Reasons to Believe events team (Bryan Rohrenbacher, Catherine Clark, and Esmeralda Cabañas) who arranged for extensive Q&A sessions in connection with my speaking opportunities. These interactions significantly augmented my research and writing, making the book more comprehensive.

Special thanks and commendations go to Drs. John Bloom, Steve Chua, Joe Miller, Fazale Rana, and Ken Wolgemuth. Their review of the manuscript resulted in clearer wording and enhanced figures. Their work and encouragement were a major benefit and blessing.

As always, Kathy (my partner in life and ministry) handled extra tasks to ensure that I had time to study and write. She also took the lead role in editing the manuscript. Hers was the challenging task of balancing my demand for scientific and theological soundness with the layreader's need for accessibility. I'm convinced her efforts propelled the book closer to this goal.

The editorial team at Reasons to Believe further improved the book's readability and thoroughness. Both Joe Aguirre and RTB Press editor-in-chief

Sandra Dimas combed through the entire manuscript, further enhancing its clarity. Helena Heredia labored diligently over the book's hundreds of endnotes. And Maureen Moser, Jocelyn King, and Karrie Cano took on a variety of important editorial tasks to bring this text to completion. Danielle Camorlinga designed the book's cover, while Anastasia Young and Stacy Nagai prepared back cover copy. Longtime friend Christine Talley designed the interior layout.

These brief acknowledgements seem woefully inadequate to convey my depth of gratitude for the effort and skills devoted by my RTB colleagues. I wish I were better at expressing how much I appreciate the efforts and skills of Reason to Believe colleagues. I am profoundly grateful to them and for them. To God be all glory, honor, and praise.

Chapter 1

A Profound Stumbling Block

The Bible is filled with stories, some more widely known and loved than others. The exodus and the nativity stand out. The story of Noah and the ark, or at least the version of the story taught in churches, Sunday school classes, and children's picture books may well be the most widely known of all. However, this story is beloved only in its fantasy version. By contrast, any hint that the story may refer to an actual historical event elicits extreme skepticism and searing derision.

In an article for a conservative Christian magazine, microbiologist Bert Thompson wrote, "It would be difficult to find an account from any period of biblical history that has been ridiculed more frequently, or with greater derisiveness, than the story of the Flood."[1] Geologist and paleontologist Donald Prothero wrote in *Skeptic* magazine, "Some of their [global flood defenders'] most egregious attempts to twist reality to fit their bizarre views are found in 'flood geology,' a concoction of strange ideas about the geological record."[2]

I placed a few brief social media posts mentioning Noah's flood as a real event in the early history of humanity and was astonished by the response. The posts, so far, have drawn 8.5 million views and 14.5 thousand questions and comments—far more engagement than any of my other posts. The brief sampling of comments (below) from Bible skeptics were ubiquitously derisive or contemptuous:

> "Anyone who actually believes the story of Noah's ark has the IQ of a cabbage."
>
> "For an otherwise sane adult to pretend that there was a biblical flood

is morally reprehensible."

"The fact that you believe ancient fairy tales negates any value your opinion might have had."

"Explain to me how penguins waddled all the way from Antarctica to the Middle East, kangaroos hopped over from Australia, and sloths managed to make it from South America."

"The ark story is the best way to get Christians to question their beliefs. It's so bizarre and ridiculous."

Most of the comments from Bible defenders either denied the factuality of the Bible's historical and scientific content or insisted that scientists are completely wrong about Earth's past history. Here are a few examples:

"The Bible is symbology and metaphor. It's not literal."

"The list of reasons why it [Noah's flood] is impossible would fill multiple pages. And the list of reasons why it was unnecessary is also extensive."

"I totally disagree with Noah's flood being a localized event. I believe it was an astral catastrophe of global consequence. It tore Pangea apart and whole mountain ranges were raised up. The flood and its causation nearly destroyed the earth."

"The Bible is about philosophy. It's not a historical or scientific resource."

What especially disturbed me were the thousands of personal insults that people lobbed at one another and at me. As one of my friends responded, "Wow! Noah's ark really triggers people!"

Rejection Based on Misperception

Sadly, what triggers this strong and derisive reaction to the story of Noah's flood is *not* the biblical account but, rather, common misunderstandings, misinterpretations, and misrepresentations of what Genesis 6–11 says. As one respondent to this torrent of mockery added, "The most skeptical and argumentative individuals who seek to impugn the Bible are those who, judging by their questions, appear to have never read it conscientiously."

Nearly every time I speak, regardless of topic, people ask me about Noah's flood. The most common question goes something like this: How can anyone rationally believe that just a few thousand years ago a flood covered Earth's entire surface, formed nearly all Earth's geological features, including nearly all Earth's fossils, and destroyed all animals except those on the ark? The assumption behind this question is obvious: a historical reading of Genesis 6–9 *compels* this interpretation of the text.

Given the overwhelming scientific evidence negating this understanding of the flood story, many people choose to dismiss the rest of the Bible as having no credence. Although I have written briefly about Noah's flood in a previous book,[3] the magnitude of this response points to the need for a more thorough treatment of the subject. The primary reason—made abundantly clear in both face-to-face and online interactions with people—is that the Genesis flood remains a huge barrier for people, an obstacle to trusting the Bible as a wholly truthful message from the one true God. A second but very important reason is that recent scientific discoveries shed new and helpful light on early human history and on geological events that are part of that history.

One contributor to the problem is the erroneous belief that Genesis tells the *whole* biblical story of Noah's flood. Even people who attend church regularly seem to think Genesis is the only book that speaks about Noah and the events of his day, but it isn't. Much relevant commentary on the flood comes from other portions of Scripture—not to mention the record of nature. Failure to incorporate this additional content into our understanding of the event can lead to a mistaken and/or incomplete interpretation of what God reveals about it. Given that 66 books comprise the Bible, we should not be surprised that not *all* the answers to questions about Noah's flood appear in Genesis.

In the twenty-first century, geology and geophysics are no longer immature or unfamiliar scientific disciplines. Nor are many of their findings obscure and inaccessible to most people. Earth's history, structure, and features have become more firmly established and broadly comprehended. Even the science of plate tectonics has become at least somewhat familiar to most people in most nations—or at least available to readers via the internet.

Adults on every continent, especially those with formal training in the sciences, recognize that no geological evidence exists—at least none confirmed as credible by the scientific research community—for a globally extensive flood during the past several thousand years. The idea that one such flood accounts for virtually all Earth's geological features and fossils, as some Christians assert, appears even more preposterous. Sadly, the perception (often perpetuated in picture Bibles and children's storybooks) that Genesis 6–9 describes a globally extensive flood that destroyed all living creatures on the planet, except those on Noah's ark, and sculpted all Earth's major geological features comes with consequences. This view has caused people around the world to dismiss the Bible—and Christianity—as implausible and, thus, of no great relevance for life today.

Within the Christian community, this distorted interpretation of the flood account (aka "flood geology") prompts some theologians and pastors to conclude that Genesis 1–11 must be borrowed mythology or, perhaps, a nonhistorical or nonliteral, possibly symbolic, text. Other outspoken Christians conclude that the community of physical scientists has banded together to deceive the public about Earth's geological history.

Bible vs. Science—a Pervasive Misperception

The idea that Christianity and established science stand in conflict has been embraced by both Christians and non-Christians in many parts of the world today. A Pew Research Center survey of self-identified Christians in the United States found that a majority (55%) see science and religion as "often in conflict."[4] Among US adults with no religious affiliation, 76% agree that science and religion are in conflict.

Among those in either group who see a clash between science and Scripture, the biblical portion most often cited is Genesis 1–11, with Genesis 6–9 (the account of Noah's flood) identified as the most problematic. The fact that young-earth creationists (YECs) and mainstream research scientists continue to engage in a "no-holds-barred" public battle only adds to this perception. What few YECs and even fewer research scientists recognize, however, is that flood geology did not emerge from YEC ideology. Rather, the two "doctrines" developed together.

As historian Ronald Numbers has documented, those who gave birth to the modern YEC movement considered flood geology essential to sustaining their belief that Earth is less than 10,000 years old.[5] During the early days of young-earth creationism, the significance of flood geology tended to eclipse that of Earth's age and the special creation of Adam and Eve (see chapter 7). Belief in a recent global flood continues to be foundational to YEC theology. As Ken Ham and Bodie Hodge of Answers in Genesis freely acknowledge, "This is really what the age of the earth debate boils down to—a global Flood or slow, gradual, millions of years."[6] Ham and Hodge add, "Secular time scales are based on the idea that there were no significant [as in 'global'] catastrophes that occurred in the past."[7]

Given that beliefs about Noah's flood reside at the core of the seeming science-faith controversy, a rigorous study of what the Bible teaches about the flood seems the most appropriate starting point for addressing and potentially settling the perceived—but wholly unnecessary—conflict between biblical and scientific history.

The Importance of Resolution

In other words, the clash between young-earth creationism and science, particularly the disciplines of astronomy, physics, and geology, cannot be resolved until the conflict between "flood geology" and these disciplines is understood and resolved. To restore confidence in the Bible as the inspired, inerrant Word of God, the biblical account of Noah and the flood that wiped out all humans and animals outside the ark must demonstrate a substantial degree of credibility. As Augustine of Hippo wrote sixteen centuries ago,

> Usually, even a non-Christian knows something about the earth, the heavens, and the other elements of this world, about the motion and orbit of the stars and even their size and relative positions . . . and so forth, and this knowledge he holds to as being certain from reason and experience. Now, it is a disgraceful and dangerous thing for an infidel to hear a Christian, presumably giving the meaning of Holy Scripture, talking nonsense on these topics; and we should take all means to prevent such an embarrassing situation, in which people show up vast ignorance in a Christian and laugh it to scorn. The shame is not so much that an ignorant individual is derided, but that people outside the household of faith think our sacred writers held such opinions, and, to the great loss of those for whose salvation we toil, the writers of our Scripture are criticized and rejected as unlearned men. If they find a Christian mistaken in a field which they themselves know well and hear him maintaining his foolish opinions about our books, how are they going to believe those books in matters concerning the resurrection of the dead, the hope of eternal life, and the kingdom of heaven, when they think their pages are full of falsehoods and

> on facts which they themselves have learnt from experience and the light of reason?[8]

As a research scientist and a Christian, I am convinced that the Bible does hold the required measure of scientific credibility. Nevertheless, many individuals (scientists and nonscientists) will choose to ignore what God reveals about himself in the realm of nature (Romans 1:18–21) and in the words of Scripture. As Jesus explained, wide is the gate and broad is the way that leads to destruction, and narrow is the gate that leads to life (Matthew 7:13–14). Not everyone is willing to examine and embrace revealed truth.

The gate that leads to a recognition of God's revelation of himself in the realm of nature is not so narrow, however, as to exclude virtually all research scientists. The apostle Paul, for example, won a number of converts among the scholars in the Areopagus in Athens (Acts 17:16–34). Christ has commissioned his followers to go and make disciples among *all* the world's people groups, including the thinkers, questioners, and learners of all cultures (Matthew 28:19–20).

For more than a tiny fraction of thoughtful adults to perceive that the Bible and science are allies and not enemies, they must know that the biblical flood story is compatible with the findings and conclusions of at least a substantial fraction of research astronomers, physicists, and geologists. They must know that a viable interpretation, one with a measure of scientific credibility, is both possible and defensible.

Making Peace vs. Keeping Peace

Many Christian scholars advocate for a "truce," an agreement by opposing parties in the Bible-science conflict to cease from vindictive language and personal attacks against one another. They call upon all participants to disagree in a spirit of kindness, gentleness, and respect so that both Christians and non-Christians will see a reflection of Jesus's words in John 13:35: "By this everyone will know that you are my disciples, if you love one another." Such a stance certainly would be better than the all-too-familiar open antagonism and vitriol. While this goal is worth pursuing, it nevertheless falls short of what's best. A higher objective is to *make* peace, not merely to *keep* the peace.

My hope for this book is that it will open a pathway to peace—a genuine resolution that can be achieved without compromising either biblical or scientific truth. I'm convinced a resolution does exist, one that requires none of the proposed redefinitions of biblical inspiration and inerrancy.[9] I believe a resolution exists that fully aligns with *all* the articles in *The Chicago Statement on Biblical Inerrancy*[10] and in *The Chicago Statement on Biblical Hermeneutics.*[11]

Though many scientists and theologians consider peaceful resolution beyond reach, I'm reminded of Hosea 4:6, which says, "My people are destroyed from lack of knowledge." For many people, a missing piece of knowledge—or a misunderstanding of the available knowledge—is the real stumbling block. When the missing knowledge or clarification is provided, people no longer feel compelled to compartmentalize what they know—to separate what they think the Bible is saying from what they think science is saying.

My aim, always, is to affirm the "two books" doctrine, codified in article 2 of the Belgic Confession.[12] This historic confession of faith affirms that God has revealed himself and his attributes through two trustworthy, utterly reliable, and harmonious revelations, the book of Scripture and the book of nature. In *Rescuing Inerrancy*, I offered brief comments on how the biblical flood account and the latest scientific data from both geology and anthropology agree.[13] However, recent experience indicates the need for a more thorough demonstration of how a historical, literal, and consistent analysis of the biblical texts on Noah's flood aligns with the latest and most firmly established science.

Chapter 2

The Basic Story

Even in parts of the world where Christianity is a minority religion, people seem to have some inkling of a biblical story about a great flood. In the United States, it's represented in children's art, children's books and Bibles, as well as in nursery decor, with sweet cartoonish drawings of lions and tigers, polar bears and penguins, giraffes and elephants, kangaroos and koalas, all marching happily, two-by-two, up a ramp onto Noah's giant wooden boat. In recent years, many people have visited the replica of Noah's ark in Kentucky, which purports to bring this story to life.

How ironic that this Bible story, most often treated as a whimsical fairy tale or as an appropriate theme park for a family outing, recounts one of the darkest and deadliest events in human history! Four entire chapters of Genesis (6–9) are devoted to recounting this grievous disaster, which is also mentioned in other parts of the Bible. A synopsis can be drawn from Genesis 6:13–17, Genesis 7:19–24, and Genesis 8:15–17:

> God said to Noah, "I am going to put an end to all people, for the earth is filled with violence because of them. . . . make yourself an ark of cypress wood [gopher wood in most English translations]. . . I am going to bring floodwaters on the earth." The waters rose and covered the mountains to a depth of more than fifteen cubits. . . . The waters flooded the earth for a hundred and fifty days. Then God said to Noah, "Come out of the ark, you and your wife and your sons and their wives. Bring out every kind of living creature that is with you."[1]

While the biblical account of this flood is understandably interpreted as a demonstration of God's judgment upon human evil and its far-reaching impact, the context of all biblical passages referring to Noah and the flood is one of rescue and recovery. The theme of the biblical story is mercy, God's mercy. The violence of Noah's contemporaries appears to have reached such a level as to threaten humanity with self-extermination (see chapter 10 in this book). Murder was rampant and out of control. Genesis 6–9 shows that God acted to prevent the extinction of human life on Earth.

To gain an accurate picture of what is happening in this story requires a close study of the original Hebrew text, including the chapters leading up to the flood and the chapters following it. These passages provide crucial context.

Another important interpretive factor is vocabulary. Too often through the centuries, Bible readers have failed to consider the variety of meanings associated with certain Hebrew words used in the account. Even a cursory reading, however, can be helpful in evaluating the plausibility of various scholars' claims to have discovered when, where, and how the biblical flood occurred.

Chapters 10–15 address multiple statements in the New Testament along with many more Old Testament texts beyond Genesis 9 that shed important interpretive light on the account of Noah's flood. Here, I present an overview of the essential elements in Genesis 4–8 that pertain to Noah's flood, giving particular attention, where necessary, to the meaning of the original Hebrew words:

1. A significant factor leading up to Noah's flood was the lack of restraint on violence and murder. Cain became the first murderer when he took the life of his brother Abel. According to Genesis 4, Cain's murder encouraged others. For example, Lamech, a descendant of Cain, boasted, "I have killed a man for wounding me, a young man for injuring me. If Cain is avenged seven times, then Lamech seventy-seven times" (Genesis 4:23–24).
2. Humanity became so wicked that people's thoughts and intentions were "only evil all the time" (Genesis 6:5). The Hebrew word used in Genesis 6:11, 13 to describe the violence expressed by humans other than Noah's family is *hamas*, a reference to extreme wickedness, violence, malicious cruelty, physical attack, or oppression.[2] This Hebrew word is used elsewhere in the Old Testament to refer to the behavior of certain individuals, but only in Genesis 6 is it used for an entire population. The evil and violence being expressed by Noah's contemporaries

was widespread, unparalleled in human history to this day. Humanity's moral corruption impacted and irreparably damaged even their animals and their land.

3. God grieved deeply over the condition of humanity before bringing his justice upon them (Genesis 6:6–7).
4. One person stood out as "a righteous man, blameless among the people of his time" (Genesis 6:9). This individual, Noah, "found favor in the eyes of the Lord" (Genesis 6:8).
5. So, God informed Noah of his plan "to put an end to all people" (Genesis 6:13), as well as the reason for it, and then instructed Noah to build a large vessel, an "ark," giving very specific construction details (Genesis 6:14–17).
6. God promised to establish his covenant with Noah and told him to bring into the ark his wife and family, pairs of certain animals (see chapter 12 for identification of which animals are included in these pairs), and sufficient food for all (Genesis 6:18–21).
7. Noah followed God's instructions (Genesis 6:22). He may have taken as long as a hundred years to build the ark.[3] During this lengthy construction period, he preached to his contemporaries, warning them of coming judgment and exhorting them to repent.[4]
8. A week before the flood was to begin, God told Noah to move into the ark with his family and animals (Genesis 7:6–10). Additional instructions included taking along "seven pairs of every kind of clean animal, a male and its mate, and one pair of every unclean animal, a male and its mate" (Genesis 7:2), to keep their "kinds alive throughout the earth" (Genesis 7:3).
9. Again, Noah followed instructions. The animals actually "came to Noah and entered the ark" (Genesis 7:15) and the Lord shut its door (Genesis 7:16).
10. On a specific date in the six hundredth year of Noah's life, "all the springs of the great deep burst forth, and the floodgates of the heavens were opened" (Genesis 7:11). The rains continued for "forty days and forty nights" (Genesis 7:12).
11. The floodwaters rose higher and higher until they covered all the visible mountains (or hills) to a depth of more than 20 feet, killing all the people and their animals outside the ark (Genesis 7:18–23).
12. God sent a wind over the earth (Genesis 8:1). The waters began to recede, and after 150 days the ark stopped drifting and came to rest on

the mountains (or hills) of Ararat (Genesis 8:3–4).

13. Roughly seven and a half months after the flood had begun, Noah could see mountains (or hills) on the horizon (Genesis 8:5).
14. After 40 more days, Noah opened the ark's window and released a raven and a dove to determine the extent of the water's recession (Genesis 8:7–8). The raven kept on flying, but the dove could find no dry ground on which to set its feet (Genesis 8:9).
15. Noah waited seven more days and again sent out the dove. This time the dove returned with a freshly plucked olive leaf in its beak (Genesis 8:10–11).
16. Noah waited seven more days and sent out the dove again. This time the dove did not return to him (Genesis 8:12).
17. Some 375–377 days after the deluge had begun, Noah removed the covering from the ark and saw that "the surface of the ground was dry" (Genesis 8:13).
18. On that same day, God told Noah to come out of the ark, bringing his family and all the animals with them (Genesis 8:14–19).

The first thing Noah did after setting his feet on dry ground was to build an altar to the Lord, offering some of the "clean" animals [animals later defined as suitable for offering to God] in sacrifice (Genesis 8:20). In this way Noah expressed gratitude to God for keeping him and his family alive through the massive devastation.

As you may know, the book of Genesis is not the only ancient text to tell of a cataclysmic flood that wiped out the whole of humanity. Several similar accounts can be found in ancient Near Eastern literature. In fact, they exist throughout the ancient world. The next chapter describes these flood accounts, their depiction of the geographical and demographic extent, and their scientific and theological implications.

Chapter 3

Everywhere a Flood Story

The Genesis story of a devastating flood that brought judgment upon humanity is the best-known account of a universal flood in the ancient world. (The term *universal* here, and in the rest of this chapter, refers to the totality of Earth's human population.) However, as mentioned in chapter 1, it's just one of many such accounts. Stories of a devastating deluge that occurred early in human history are part of the myths and legends of virtually every ancient culture, every ethnic group. These stories exist even among people groups who have lived for millennia in deserts and on high mountain ranges. They are found among people groups currently living in Africa, Asia, Australia, Europe, Iceland, New Guinea, North America, Polynesia, and South America.

Flood Stories Abound

Twentieth century German astronomer Johannes Karl Richard Riem, in his book *Die Sintflut in Sage und Wissenschaft* (The Flood in Legend and Science), documented nearly 300 distinct flood legends from across the world's cultures. In each one, all but a select few humans were wiped out.[1] French Hellenist, Assyriologist, and archaeologist François Lenormant described over 200 ancient flood legends in his book *Les Origines de l'Histoire d'apres la Bible: de la Création de l'Homme au Déluge* (The Origins of History from the Bible: From the Creation of Man to the Flood).[2] Scottish social anthropologist Sir James George Frazer, German ethnographic scholar Richard Andree, and German anthropologist and geophysicist Georg Gerland each described, in their respective books, over 100 ancient creation myths (which included references to a great flood) with some, though by no means total, overlap.[3] Alan Dundes, professor of anthropology and folklore at the University of California,

Berkeley, edited the book *The Flood Myth*, in which 12 scholars each wrote chapters describing ancient legends of a universal flood among different ethnic people groups living in the Middle East, Greece, Mesoamerica, South America, Australia, Cameroon, the Philippines, and the Indian subcontinent, including Sri Lanka.[4]

Civil engineer, historian, and global flood proponent Nick Liguori, in his lengthy book *Echoes of Ararat*, claims that flood accounts are more prevalent than previously thought in the oral traditions of North and South American native tribes.[5] He maintains that more than 300 such stories exist. Before publication of Liguori's book, biochemist Duane Gish and engineering geologist John Morris of the Institute for Creation Research wrote that they had assembled 270 accounts of a universal flood from historical and anthropological literature and 200 from missionaries who worked among the ancient people groups of the world.[6]

Historians and scientists of the past two centuries are not the only ones who make reference to the pervasiveness of flood stories. The Jewish historian Flavius Josephus made this comment in the first century AD:

> Now all writers of the barbarian histories make mention of this flood, and of this ark; among whom is Berosus the Chaldean . . . . Hieronymous the Egyptian also, who wrote the Phoenician Antiquities, and Mnaseas, and a great many more, make mention of the same.[7]

John Woodward, a British naturalist, geologist, and physician wrote in 1695,

> The unanimous Tradition of an Universal Deluge amongst all the most ancient Gentile nations: particularly the Scythians, the Persians, the Babylonians, the Bithynians, Phrygians, Lydians, Cilicians, and other People of Asia Minor: the Hierapolitans, Phonicians, and other Inhabitants of Syria: the Eygptians, Carthagians, and other African Nations: the ancient Inhabitants of the several Parts of Greece: and of the other Countries of Europe: the old Germans: the Gauls: the Romans: the ancient Inhabitants of Spain, and even the Britains themselves, the first inhabitants of this Island: Proving that the great Devastation and Havock the Deluge made, both

> of the Earth itself, of the Generality of Mankind, of Brutes, and all Animals, had wrought a deep and very sensible Impression upon the Minds of these ancient Nations.[8]

Ancient universal flood legends are found among ethnic people groups of Africa, Asia, Australia, Europe, North America, Polynesia, and South America. Among only a few groups in Africa and Asia have no such stories yet been uncovered.[9]

Common Elements

What all the ancient flood legends and myths share in common with each other and with the Genesis account is that long ago in human history a single catastrophic flood wiped out all people other than a special few. In nearly all the stories, the only people to survive the flood are those closely related to each other. The surviving family in most cases is described as virtuous or righteous in contrast to the people who drowned. In about three quarters of the stories, survival is by means of a floating vessel, and a postflood rainbow is mentioned. In most of the stories, the flood comes in response to pervasive and pernicious evil, evil people are forewarned that judgment is coming, and certain kinds of animals are included in the rescue. Other common elements are the mention of geographical place names, divine judgment as the cause of the calamitous flood, the floating vessel eventually landing on a hill or mountain, and birds being sent out to test whether the floodwaters have subsided. In about half the stories, divine judgment is stated as the cause of the calamity. Listed in table 3.1 are the percentages of known ancient flood stories and myths that share story elements in common.

In most cases where a floating vessel is mentioned, descriptions make note of its size and structure. However, only in the Bible is the vessel's description consistent with a ship capable of remaining seaworthy and able to sustain people and their animals for a whole year.

Unlike the Bible, the lengthier and more detailed flood legends and myths (see chapter 9) seem motivated by the desire to justify a particular political rule or a two-tiered social system of nobles and serfs, where the serfs are compelled to serve, submit to, and financially support the nobility. Also, these lengthier flood legends, distinct from the Bible, reflect a polytheistic belief in which many distinct gods are struggling with one another and engaging in immoral impulses and actions.

Table 3.1 Common Story Elements in Ancient Flood Stories/Myths

story element	percent of stories
floodwater depth of over 100 meters	99
flood survivors closely related	90
geographical place names mentioned	80
survival by means of a floating vessel	75
postflood rainbow	75
flood due to pernicious, pervasive human evil	67
evil people forewarned of coming judgment	67
animals included in judgment	67
certain animals included in rescue	67
divine judgment as cause of flood calamity	50
rescue vessel comes to rest on a point of high ground	50
rainfall as a major source of floodwaters	50
rising sea level as a major source of floodwaters	40
sacrifices offered by human survivors	10
number of human survivors = 8	10

Implications of Widespread Flood Legends

Today's historians and anthropologists seem to agree that the only rational explanation for the pervasive and parallel flood legends among ancient ethnic people groups is that they all experienced, at some time in their past, a devastating flood that wiped out all but a few survivors. The only *ongoing* debate is whether multiple catastrophic floods occurred in different regions at approximately the same time in human history or just one flood. A second way the debate is framed is in consideration of two questions: Do the similarities among the several hundred ancient flood stories serve as evidence for cultural transmission of a single event from the distant past before human migration and colonization got underway? Or, do they serve as an indication that independent comparable experiences occurred at roughly similar dates in the distant past?

Many scholars have pointed out noticeable trends in the diversity of ancient flood legends. One such trend is that the greater the geographical distance from the Middle and Near East, the greater the differences from the account recorded in Genesis 6–11.[10] A second trend relates to the scientific credibility of the ancient flood account. The only one free of scientifically implausible elements is the one in Genesis (see chapters 10–14). The greater the story's geographical distance from the Near and Middle East, the more scientifically absurd elements it includes. A third trend is a linguistic one. The greater the linguistic distance from the ancient languages spoken in the Near and Middle East, the more widely the legend's specifics differ from those of the Genesis flood account.

The trend line of geographical and linguistic differences among ancient flood legends appears most consistent with the cultural transmission of a single event that occurred before human migration and colonization began (see chapters 10–12 and Genesis 10–11) and before the diversification of human languages (see Genesis 11:7–9). Scholars who have compared more than 200 ancient flood legends and myths say that these stories share far too much in common to be reports of multiple independent flood events occurring at different times. Charles Martin, an expert in comparative literature, devoted an entire book to establishing this conclusion.[11]

Another scholar, Bryon Nelson, a Lutheran pastor who studied zoology and genetics at Rutgers University, observes in his book *Deluge Story in Stone*:

> The only alternative to the theory that there was one single flood catastrophe, universal as far as the human race is concerned, which is the basis of all the Deluge legends, is that there were many separate local floods each of which is the basis of a different flood account. This alternative has this chief objection: namely, that the remarkable similarity in the Deluge legends is not accounted for.[12]

Francis Schaeffer, the eminent twentieth-century philosopher, theologian, and author draws this conclusion in *Genesis in Space and Time*:

> Among the common myths in the world's history, no other one is so widespread as the story of the flood. From China to the American Indians and even the pre-Colombian Indians, one finds in strange forms the myth of the great flood. Most of these myths have weird elements—foolish elements, for example the descriptions of the boat that was used. In the Bible these strange and foolish elements are not there. We would say, then, that the Bible gives us the history of the flood; the myths all over the world are contorted but show that men everywhere have a memory of it. Here in the Bible is the one flood story whose details, including the construction of the vessel, are reasonable.[13]

The fact that legends or myths remarkably similar to the Genesis account of Noah's flood appear among nearly all the world's ancient peoples and cultures stands as striking historical evidence for a single catastrophic flood that impacted all of humanity. That these flood legends and myths become gradually more diverse with linguistic and geographical distance from the ancient Near and Middle East provides historical evidence that humans at some time after Noah's flood did, indeed, disperse from a single region, just as Genesis 10–11 declares. Genesis 10–11 says that God scattered humanity from one region or a few contiguous regions into all the habitable continents and islands of the world.

With the story of a catastrophic flood in the lore and/or literature of nearly all ancient societies, several geologists and archaeologists have endeavored to identify a discoverable past flood that might correspond to these stories. The next chapter examines their efforts and conclusions.

Which Flood is *the* Flood?

For centuries, Christian, Jewish, and Muslim scholars have searched for physical evidence of Noah's flood, the account presented in their holy books. Not until the twentieth century did any evidence that seemed worth investigation catch the attention of science-trained scholars. Before long, a flood of hypotheses arose.

Mesopotamian Flood Deposits

In 1929, famed British archaeologist Sir Charles Leonard Woolley announced his discovery of a flood deposit near the ruins of Ur in southern Mesopotamia. He identified this deposit as having been laid down during Noah's flood.[1] That same year, archaeologists Stephen Herbert Langdon and L. Charles Watelin found smaller flood deposits near the ruins of Kish in central Mesopotamia.[2] Initially, archaeologists thought the flood deposits near Ur and Kish were all part of one huge flood deposit that covered all of Mesopotamia. Subsequent research, however, showed that none of the deposits covered an area that large, and the deposit near Ur was shown to be 900 years older than the deposits near Kish.

In 1931, archaeologist Erich Schmidt discovered a flood stratum near Shuruppak, an ancient city located about halfway between Ur and Kish.[3] This stratum proved to be unrelated to the deposit near Ur, as well as to two, if not all three, deposits found near Kish.

By the beginning of the second half of the twentieth century, archaeologists had been able to determine that the previously discovered Mesopotamian flood deposits could not have been related to Noah's flood. Rather, they were the result of occasional local torrential rains and occasional overflowing of the

Tigris and Euphrates rivers. What's more, with dates ranging from 3500–2600 BC, these deposits are much too recent. They certainly did not predate a post-flood scattering of humanity from the Near and Middle East into all of Africa, Asia, and Europe, as described in Genesis 10–11.

Black Sea Inundation

In 1997, geologists William Ryan and Walter Pitman III and their colleagues published evidence suggesting that the Black Sea had transformed from a relatively small freshwater lake into a huge salty sea. The proposed cause: a sudden inundation of Mediterranean Sea water due to rapidly melting ice from the latest ice age that dramatically raised sea levels.[4]

A year later, the same two authors published a book in which they postulated that the breaking of a land bridge joining Greece to Anatolia (the Asiatic part of Turkey) accounted for Noah's flood.[5] They claimed that this bridge, which once separated the Black Sea from the Mediterranean Sea, gave way about 8,000 years ago, leading to the formation of the Bosporus Strait. They cite fossil and geological evidence indicating that the Bosporus Strait formed about 7,600 years ago.[6] According to their scenario, the breaking of the land bridge resulted in a vast rush of water from the Mediterranean Sea into the Black Sea, nearly doubling its area.

According to their hypothesis, hundreds of feet of water would have covered 200,000 square kilometers (75,000 square miles) and raised the level of the Black Sea by nearly 100 meters (330 feet). However, stratigraphic and paleogeomorphologic data, especially carbon-14 dates for Danube delta strata and mollusk fossils, say otherwise, as geologists Liviu Giosan, Florin Filip, and Stefan Constatinescu explain.[7] Their studies of the Danube delta, which flows into the Black Sea, constrain the rise of the Black Sea to no more than 30 meters (98 feet). They conclude that "the sea level increase and the flooded area during the reconnection [of these waters] were significantly smaller than previously proposed."[8] A recently published review notes that a persistent outflow from the Black Sea to the Mediterranean has occurred throughout the past 12,000 years.[9]

In a comprehensive review of published papers on the Black Sea inundation, two geologists at Memorial University of Newfoundland concluded that the "reconnection of the Black Sea with the eastern Mediterranean occurred in a gradual fashion."[10] They added, "There not only is evidence from each sea to contradict the Flood Hypothesis, but the evidence in favour of the Outflow Hypothesis [water flow from the Black Sea to the Mediterranean] forms a single coherent story and timeline all along the gateway and its environs."[11] Consequently, the Black Sea level rise would have amounted to, at most, only a fraction of a meter or a foot per year. People living on the edge of the Black Sea would have had plenty of time to pack up their animals and belongings and move to higher ground.

The claim that the Black Sea inundation is caused by Noah's flood faces another challenge. The inundation covered far too small an area to wipe out all humans and their animals living at that time. By 7,600 years ago, humans already had dispersed from the Near and Middle East and eastern Africa to occupy all of Africa, Asia, Australia, North America, and South America. So Noah's flood must have occurred much earlier than 8,000 years ago (see chapter 11).

Indian Ocean Asteroid Strike

Environmental archaeologist William Bruce Masse, of the Los Alamos National Laboratory, has interpreted flood myths throughout the world (see chapter 3) and examined petroglyphs and certain chevrons (ocean-floor deposits) along the African coast. He hypothesizes that a 5-kilometer-diameter comet struck the Indian Ocean around 2800 BC. Such an event would have generated 180-meter-high (600-foot-high) megatsunamis and sent them crashing into all the world's coastlines.[12] Although he didn't explicitly endorse this asteroid strike as an explanation for Noah's flood, many Christians who became aware of his research and hypothesis did.

Most scientists, however, express doubt about Masse's comet strike hypothesis. The chevron data Masse cites appears limited and inconclusive. Given the claimed recency of the strike, the chevron data supporting the strike *should* (in the scientific sense) be pervasive and dramatic, but it isn't. Also, the impact of a 5-kilometer-diameter comet would have been far more catastrophic than the impact of Noah's flood as described in Genesis 6–9. Had it struck, the whole planet would have experienced a severe winter lasting several years, and Noah's ark likely would have been smashed to smithereens.

Ice Age Floods

Since the beginning of the human era, the most dramatic flooding of Earth would have occurred during the last ice age, which lasted from 120,000 to 12,000 years ago. The latest glacial maximum (when glaciers and ice sheets reached their largest extent) ended a mere 18,000 years ago. Interglacial temperate conditions began 16,000–15,000 years ago. The present warm Holocene era began 12,000 years ago. Evidence shows that on several occasions during the previous ice age, huge amounts of continental ice sheets melted. A few of these episodes caused sea levels to rise, in a spatially variable manner, by as much as 100 meters (330 feet) within the time span of just several millennia. Each of these events flooded several million square kilometers of land.

The two most rapid and widespread flows of melted continental ice into Earth's oceans where geologists possess both accurate geographical and historical measurements occurred 14,600 and 11,500 years ago. The earlier event caused a sea level rise of about 20 meters (66 feet) over a period of less than 340 years.[13] The later event caused a sea level rise of about 18 meters (60 feet) over a 300-year period.[14] These two events flooded large coastal areas, with the greatest known devastation occurring in the southeast Asia Sunda region (see figure 11.2).

The former event also expanded Lake Agassiz, an enormous glacial lake formed by the melting of ice caps once covering Canada, Minnesota, and North Dakota (see figure 4.1). Another ice melt event occurred 8,500 years ago. It resulted in the final drainage of Lake Agassiz but caused only a minor sea level rise, compared to the two previous events. None of these melt events would have been sufficiently catastrophic to wipe out all humans and their animals living at that time.

Glaciologists conclude that two major drainages of Lake Agassiz into the Arctic and Atlantic Oceans were largely responsible for two major cooling events—"the Younger Dryas" and "the Northgrippian."[15] Each of these drainages resulted in a global sea level rise of 0.5–1.4 meters (1.6–4.6 feet) over a time span that may have been as brief as six months. Geologists Chris Turney and Heidi Brown proposed that the final drainage of Lake Agassiz 8,500 years ago could account for Noah's flood.[16] However, a flood rise of less than 1.5 meters (5 feet) wouldn't have wiped out even a tiny fraction of the humans and their animals of that time.

Flood Like No Other in Human History

How plausible are the proposals by Ryan and Pitman and Turney and Brown for identifying the biblical deluge? Although the two flood events they describe covered a great area and caused great devastation, neither reached the magnitude of the flood depicted in Genesis 6–9. In fact, they pale in comparison.

According to Genesis 7, the waters of Noah's flood rose and spread so fast that no one, not even the most fleet-footed and fit of humanity's animal companions, could have outrun it or escaped it. In Ryan and Pitman's flood

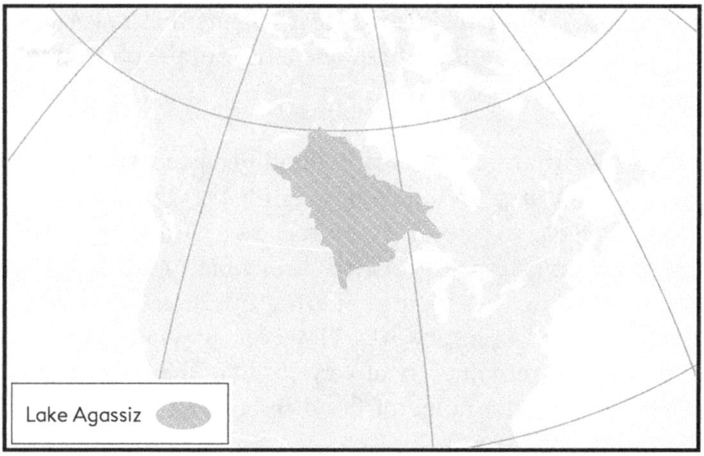

Figure 4.1: Lake Agassiz Before the Beginning of the Younger Dryas Cooling Event
Background image credit: NASA

scenario, the maximum rate of water level rise amounted to no more than one foot per day.[17] Such a rate would have been fearsome but not difficult to outdistance. The much wider inundation from drainages of Lake Agassiz raised oceanic water levels no more than 0.3 centimeters (0.12 inches) per day. Human migration away from rising sea levels would have been inconvenient, but hardly life-threatening.

According to Genesis 6–9, Noah's flood was so rapid, widespread, and devastating that neither humans nor animals associated with humans could escape it—other than those in the ark. Genesis 7:18–20 says that the surging floodwaters extended as far as the eye could see. All the land visible to Noah from the highest point on the ark was covered by at least a 15-cubit (22-foot) depth of water (Genesis 7:19–20). That's enough water to submerge a 2- to 3-story house!

Genesis 9:8–17 explicitly states that Noah's flood was an unprecedented, unrepeated event in human history. God promised "[N]ever again will all life be destroyed by the waters of a flood" (verse 11) and "[N]ever again will the waters become a flood to destroy all life" (verse 15). God's promise doesn't guarantee that devastating floods would never again cause trauma and suffering for many people. Nevertheless, the text affirms that never again would a flood wipe out all humanity other than one family of people and their animals.

Long Debate

Debates among Christians about Noah's flood persisted throughout the twentieth century and continue up to this day. While these debates have long been contentious, they seem to have become more so with every passing decade. However, they certainly aren't new. Christian scholars and church leaders have been commenting on and debating differing interpretations of Noah's flood since at least the second century AD. However, previous to the seventeenth century, these debates remained relatively cordial. The next chapter describes the earlier debates and the range of flood interpretations held by Christians prior to the seventeenth century.

Chapter 5

Before the Scientific Revolution

Debate over how to interpret the biblical account of the flood appears in the earliest writings of several Christian scholars. However, not until the eighteenth-century advent of modern geological science did the Genesis flood become a controversy dividing Christians and pitting scientists against Bible-believers. Not until the twentieth century did the controversy heat up to a degree that caused many people, both inside and outside the church, to conclude that the Bible is both inaccurate and irrelevant—a book that cannot be trusted to reveal meaningful truth about science or history.

While Christian scholars prior to the Scientific Revolution held differing views on Noah's flood, the range of disagreement was far narrower than it is today. Perhaps even more significantly, the demeanor they showed in discussing the topic with one another appeared far more charitable than what we see in such discussions today. My hope is that a brief review of the history of Christian commentary on Noah's flood holds value for lowering the pitch of current ongoing debates.

What especially necessitates such a review is the degree to which the flood interpretations by Christian scholars of past centuries have been distorted by those teaching and writing today. For example, in a recent public debate, Terry Mortenson of Answers in Genesis claimed, "For the first 1800 years of the church all Christian scholars believed in a global flood and a young earth."[1] This false claim has been consistently made by young-earth creationist leaders throughout the past four decades.[2] For this reason, in the next three chapters I cite original source writings of the leading Christian scholars of the first eighteen centuries of the Christian church, endeavoring to provide context for what these early scholars actually wrote about Noah's flood.

Early Church Fathers' Interpretations

"Early church fathers" refers to prominent theologians, bishops, and pastors who lived during the first few centuries after Jesus's resurrection and whose writings have survived to the present. These scholars offered various interpretations of Noah's flood at a time long before geology became an established scientific endeavor. Thus, they wrote about the Genesis account of Noah's flood primarily from a theological and moral standpoint, not from a scientific one. Here are 12 examples of the most widely known early interpreters' perspectives on Noah's flood:

1. Justin Martyr (c. 100–165 AD) considered Noah and his family and the deluge as symbolic, "wherein Christ appeared when He rose from the dead."[3] Martyr believed the flood was a historical event. However, the warning that floodwaters would cover "the whole earth," he asserted, "was not spoken to the land but to the people who obeyed Him."[4] He regarded the water, the ark's wood, and Noah's faith as symbols for baptism, the cross, and Jesus Christ.

2. Theophilus (c. 115–185 AD) affirmed that Moses authored the Genesis account of Noah's flood.[5] Like Martyr, he believed the flood was a historical event. He cited Plato's and other Greek scholars' assertions that a great deluge once impacted all humans. He pointed out, though, that these scholars were mistaken in supposing that the waters submerged only the plains and that a few people escaped by fleeing to the mountains. In his words, "The water overtopped every high hill 15 cubits [roughly 7 meters or 22 feet]" and only those who were protected in the ark were saved.[6] Theophilus went on to describe Noah's ministry during the building of the ark. In his words, Noah exhorted his contemporaries to "come thither, God calls you to repentance."[7] He stated further that at the time of Noah humanity was constrained to "the land of Arabia and Chaldaea" and spoke just one language. Then, after the flood, humans spread to all Earth's landmasses and developed many distinct languages.[8]

3. Irenaeus (c. 130–202 AD) viewed Noah's flood as an example, along with the destruction of Sodom and Gomorrah, of God's judgment against reprobate societies and God's mercy toward the righteous. He saw Noah's flood as a historical event that brought divine judgment upon a universally corrupt society of humans.[9] He used God's salvation of Noah and his family as an illustration of the continuity of God's redemptive plan for humanity, a plan culminating in the salvation of believers through the sacrifice of Jesus Christ.

4. Clement of Alexandria (c. 150–215 AD), like Theophilus, noted that for several centuries Greek scholars had spoken about an ancient flood that impacted all humans on Earth.[10] Also, like Theophilus, he wrote that "Noah preached repentance."[11] Clement viewed the Genesis story of the flood as historical in nature, providing moral and allegorical lessons for humans living at all times.

5. Tertullian (c. 155–225 AD), in his dissertation *On the Apparel of Women*, referred to Noah as "the survivor of the deluge" that was a "world-wide calamity."[12] In his treatise *On the Pallium*, Tertullian makes no mention of Noah or his ark. In reference to Earth, he describes "a time when her whole orb, withal, underwent mutation, overrun by all waters."[13] In the context of describing creation, he considered seashells found on mountains as evidence that those mountains were at one time under water. The creation context suggests that Tertullian was referring to Earth before creation day three, when landmasses had not yet risen above sea level (Genesis 1:9–10), a point he made clear in *The Treatise Against Hermogenes*.[14]

6. Origen (c. 184–253 AD), in chapters 58 and 59 of *Origen Against Celsus*, Book VI, aimed to address Celsus's errors in interpreting Genesis 6 and 7. He claimed that Celsus assigned much too broad a definition to "world," as used in these texts. Twice Origen explained that the world of those who perished in the deluge should be understood as "regions on earth."[15] He added, "In many parts of Scripture it is usual to call the 'world'" that region of human "existence upon earth."[16] In his *Homilies on Genesis*, Origen briefly discussed Noah's flood.[17] He accepted Genesis 6–9 as a historical account and argued that readers of Genesis needed to regard the events of Noah's flood as symbolic of baptism and purification. He believed Noah's flood represented a purification of the human soul from sin.

7. Gregory of Nyssa (c. 335–394 AD) wrote that God entrusted Noah "with saving the whole world, or rather the seeds of the world, in a small ark which escaped the deluge."[18] He gave no indication of his belief that the flood was anything other than a historical event.

8. John Chrysostom (c. 347–407 AD), in *Homilies on Genesis*, considered the account of Noah's flood an example of "the precision of Scripture."[19] He endorsed a literal interpretation of Genesis 6–9. For example, he gave credence to Noah's having lived an actual 500 years before starting to build the ark and taking another hundred years to complete it.[20] He also noted that after the flood, God shortened the maximum human lifespan down to about 120 years.[21] Chrysostom affirmed that Noah preached to his contemporaries during the ark's construction, calling them to repent.[22] He interpreted Genesis 6–9 as stating, literally, that all humans outside the ark "were wiped off the earth"[23] and that the floodwaters "submerged all the mountains."[24] Chrysostom referred to Noah's flood as God's rescuing humanity, washing and cleansing the human race from corruption and wickedness.[25]

9. Jerome of Stridon (c. 344–420 AD), best known for his Latin translation of the Bible (later known as the Vulgate), briefly addressed Noah's flood in one of his letters and in his book, *Hebrew Questions on Genesis*. In Letter 51, he refuted speculation that the Genesis 1 mention of "waters above" referred to angelic beings by explaining that in Genesis 7, water (not angels) came from the windows of heaven above and the fountains of the deep below to cover "the whole earth."[26] Jerome's argument indicates his belief that Noah's flood was a literal, historical event.

Jerome does not clarify, here, what he meant by "the whole earth." In his translation of the Bible, though, he would have observed that in multiple passages, the context makes clear that something less than the entire surface of planet Earth is implied by "the whole world" or "the whole earth" (see chapter 10). In *Hebrew Questions on Genesis*, Jerome commented specifically on Genesis 6:14, 16, and Genesis 8:6–7.[27] In describing the details of the ark's door, window, and sealing pitch, he made clear his understanding of the flood account as history.

10. Augustine of Hippo (354–430 AD) in his monumental work, *The City of God*, devoted a chapter to Noah's flood.[28] He argued that to accept Genesis 6–9 as an account only of true historical events and, at the same time, to reject its allegorical significance would be a mistake. In other words, Augustine approached Genesis 6–9 as both history *and* allegory.

Augustine wrote of the ark as a ship of great size that rescued Noah's family (plus a large population of animals) and as a symbol of the Christian church. He pointed out the lack of any need for Noah to bring on board animals that can live in the water, such as fish and seabirds. While noting that food stores for the herbivore animals wouldn't be a problem, he saw that Noah would need to ensure that there was adequate food—whether live prey, preserved meat, and/or processed vegetables and grains—for the carnivores' survival. He saw Noah and his family as representatives of righteous humans saved by God's mercy and grace.

Augustine also wrote about Noah's flood in his *Reply to Faustus the Manichaean*.[29] There he drew several analogies between the historical event of the flood and the gospel of salvation through Jesus Christ. In his words, "Noah with his family is saved by water and wood, as the family of Christ is saved by baptism."[30] The "ark is made of beams . . . as the Church is constructed of saints prepared for every good work."[31] He saw the pitch that fastened together the ark's beams as a sign of Christians' love, "which keeps the brotherly connection from being impaired,"[32] and the inclusion of all kinds of animals in the ark as signifying that "the Church contains all nations."[33]

11. Cyril of Alexandria (376–444 AD), in his commentary on the Gospel of John, referred briefly to Noah's flood. He noted the important distinction in 2 Peter 2:5 that "the Flood [was] brought in upon the world of the ungodly."[34] He affirmed that because of Noah's faith in God, God preserved him.

12. Theodoret of Cyrus (393–458 AD) wrote about Noah's flood in his *Questions on Genesis* and his *Questions and Answers to the Orthodox*. He considered the account of the flood in Genesis to be literal and historical.[35] He, too, believed the Genesis account of the flood held moral and spiritual lessons for all humans of all times.

These 12 leaders of the early Christian church account for nearly all the writings about Noah's flood from that early era. Although their interpretations varied from historical to allegorical, moral, or typological, they generally accepted the flood as an actual event. Given the lack of physical and geological knowledge about Earth's history available to them, it seems unsurprising that their views on the flood were predominantly shaped by how they interpreted the grand theological themes of the Bible.

What may seem most striking about the writings of these 12 scholars is the complete lack of hostility, caustic language, or polemic among them, whatever their interpretive differences were. We find no evidence that any of them were hostile toward one another over their various views on the flood. What's more, although they didn't back away from polemical writings against non-Christian scholars, their polemics remained consistently polite and respectful.

As Origen noted in his writings, "world" may not have meant the globe. Twice he wrote that the world of those who perished in Noah's flood should be understood as "regions on earth." It seems worth noting also that Cyril, Archbishop of Alexandria, when quoting from 2 Peter 2:5, qualified the "world" that was flooded as "the world of the ungodly." Mortenson conflated "worldwide" with global. None of the 12 luminaries of the early church ever stated that Noah's flood covered Earth's entire surface.

While virtually all twenty-first century people possess a global perspective, people living one and two millennia ago viewed the world through a regional or cultural perspective (see chapter 10). Therefore, it seems reasonable to suggest that "world" for many early Christians meant something other than what it typically (even automatically) has meant to later readers.

Christian Scholars of the Middle Ages

During the Middle Ages, Christian scholars—building upon the writings of the early church fathers—continued to write about Noah's flood. They commented on the flood from a variety of perspectives: scientific, historical, theological, allegorical, and moral. Here are the five most frequently cited examples of flood commentary from that era:

1. Saint Bede the Venerable (673–735) touched on Noah's flood in his teaching on Genesis 1.[36] A theme throughout his writing is that the biblical text contains both a literal and a figurative message.[37] He saw the evening of the first creation day as a foreboding of the great iniquity that would befall the human race "to such an extent that all flesh deserved to be destroyed by the flood, except those in the ark."[38] He believed the waters of the second creation day foreshadowed "the ark, in which the remains of the human race and the seed of succeeding ages, so to speak, was preserved."[39] The evening of the second creation day, he wrote, referred allegorically to "the nations, forgetting the closeness of the anger or mercy of God" and meeting "to build a tower of pride."[40] Bede appears to have viewed the flood, Noah, the ark, and the subsequent rebellion against God at the tower of Babel as literal, historical events.

2. Petrus Comestor, also called Pierre le Mangeur (c. 1100–1178), a French theologian and biblical scholar, authored *Historia Scholastica*, his commentary on the writings of the Scholastics (as church scholars of that period are sometimes referred to). His book devotes three chapters to an analysis of the Genesis account of Noah and the flood.[41] Like Augustine, he viewed Genesis 6–9 as a truthful account of historical events that also includes allegorical references to salvation, God's grace, and the church.

3. Vincent of Beauvais (c. 1190–1264), a French Dominican friar, is noted for compiling an extensive theological encyclopedia, *Speculum Maius* (The Great Mirror). One of these volumes, *Speculum Historiale*,[42] includes an extensive entry on Noah's flood. While citing the historical, allegorical, and moral interpretations of the flood held by the patristics, he maintained his stance as an encyclopedia compiler, declining to state his own views.

4. Thomas Aquinas (1225–1274), in his *Summa Theologica*, asserted that the literal interpretation of the Bible is always paramount and that any spiritual or allegorical interpretation of a biblical text presupposes the literal interpretation and is built upon it. According to Aquinas, "In Holy Writ no confusion results, for all the senses [interpretations] are founded on one—the literal—from which alone can any argument be drawn."[43] He applied this principle to conclude that the story of Noah and the flood must be taken literally and historically, and not just as a parable. In the prologue to his *Commentary of Job*, Aquinas left no doubt that he considered Noah to be a real historical figure. He argued, "Clearly Noah and Daniel really were men in the nature of things."[44]

In *Summa Theologica*, Aquinas acknowledged Noah's flood as a historical event pointing to a deeper spiritual allegory. He wrote, "By the deluge He [God] destroyed from the face of the earth man whom He had made."[45] "An ark of wood preserved the human race from waters of the Deluge,"[46] which was "a figure of our Baptism, in respect of the salvation of the faithful in the Church, since then 'a few . . . souls were saved in the ark.'"[47]

5. Nicholas of Lyra (c. 1270–1349) composed a lengthy commentary on the Bible, *Postilla litteralis in Vetus et Novum Testamentum* [Literal Commentary on the Old and New Testament], in which he defended Genesis 6–9 as a historical account of Noah's flood.[48] He discussed divine reasons for the flood, the construction of the ark, and Noah's faithfulness to God as well as to the mission God had given him. He claimed that the flood account in Genesis also included symbolic features pointing to the salvation that Christ would offer to rescue us from sin.

Christian Scholars of the Reformation

Many Reformation scholars wrote about and/or preached sermons on Noah's flood. The paragraphs that follow provide an overview of flood teaching by the two most widely known and cited Reformation scholars.

1. Martin Luther (1483–1546), the seminal figure of the Protestant Reformation, wrote a lengthy commentary on the book of Genesis in which he devoted considerable attention to the account of Noah and the flood.[49] He presented Noah as a historical figure, especially highlighting Noah's role as a prophet and preacher of righteousness—one who likely endured ceaseless ridicule, insults, and abuse from the ungodly people surrounding him.[50]

In chapter 6, paragraphs 212 and 213 of his commentary on Genesis, Luther noted that the animals destroyed in the flood were those that belonged to wicked people. He wrote that "the animals perished, not because they had sinned, but because God wanted man to perish amid all his earthly possessions."[51]

The reformer went on to describe in detail the extreme debauchery of Noah's contemporaries. He explained the meaning of *hamas*, the Hebrew word used in Genesis 6 to describe the horrific degradation of human society. He wrote "[T]he correct significance of the word *hamas* is violent force, wrong, with the suspension of all law and equity, a condition where pleasure is law and everything is done not by right, but by might."[52] The result, he said, was pervasive "violence and shameful living"[53] that led to Noah's contemporaries judging "Noah's teaching and practise [sic] as wholly wrong, and their own as altogether holy."[54] The extent of wickedness seemed clear to Luther, based on the words of Genesis. He wrote, "All men were wicked, lived in idolatry and false religion and hated true worship of God. They despised the promise of the seed, and persecuted Noah."[55]

Luther asserted that God destroyed "all flesh and all the earth"[56] because "the earth was filled with violence."[57] He argued that every part of Earth impacted by human violence had to be destroyed, just as every part of the land occupied by the Sodomites had to be destroyed. "God, because of sin," Luther wrote, "has brought punishment to bear, not alone upon man, but also upon his property and domain."[58] With reference to the geographical extent of the punishment, Luther draws an analogy from his own time. In his outcry against the Roman Catholic Church, he describes that church as "a horrible abomination that fills the whole earth."[59] He then poses a question, "Does it not become clear to every man that God is compelled, as it were, to punish, yea to overturn Germany?"[60]

Nowhere in Luther's writing about Noah's flood did he explicitly state that the flood covered the entire surface of planet Earth. In fact, he explicitly states that the flood covered less than Earth's entire surface. In his words, "the greater part of the world" perished in the deluge.[61] However, he was unequivocal in stating that, apart from Noah and his family, "all mankind perished."[62]

Luther expressed his belief that the description of the ark in Genesis 7 was literal and accurate in its dimensions and details.[63] He also lent credence to Augustine's view that the ark and the floodwaters were also symbolic, representing the church and baptism.[64] He accepted the validity of two huge sources of water for the flood: torrential rains and the upwelling of underground aquifers.[65]

2. John Calvin (1509–1564), the leading French Protestant reformer and most influential person among the second generation of Protestant reformers, wrote several commentaries in addition to his seminal *Institutes of the Christian Religion*. In his extensive commentary on Genesis, Calvin, like Luther, identified Moses as the author of the book and acknowledged the story of Noah, the ark, and the flood as literal history. He agreed with Luther that all humans at that time, other than Noah and his family, had degraded into incurable reprobation. Calvin described "the deplorable wickedness," "incurable obstinacy," and "degenerate and corrupt nature" of Noah's generation.[66] He added that "wickedness was too deeply seated in their [the people's] hearts to leave any hope of repentance," that the wickedness was "continual," and that it "grew worse and worse as time advanced."[67]

Calvin noted that the wickedness of Noah's contemporaries had damaged their animals, as well. Because these creatures were "in subjection to man when he fell, they were drawn with him into the same destruction."[68]

The French theologian also explicitly stated his view that Moses's description of the ark and its dimensions was literal.[69] He declined to speculate whether the "short cubit" or "long cubit" defined the ark's size, but he accepted that the ark was of the appropriate size to accommodate all the animals that God intended for Noah to bring with him and his family.[70]

Calvin wrote that "the whole world was immersed in the waters" of the flood.[71] However, he never explicitly equated "the whole world" with the globe, or with the entire surface of Earth. More likely, he used the term to refer to the world of ungodly people, as he did in his commentary on 2 Peter 2:5 wherein he contrasts the old world of the human race with the new world of the human race.[72]

In response to the apostle Peter's words about the flood of Noah's time, Calvin commented on the enormity of "the crimes of men by which God was induced to destroy the whole world."[73] In the same paragraph, he referred to the world "as it was then," so "full of unbelievers."[74]

The reformer briefly commented on claims made by Josephus (c. 37–100 AD) and Jerome (c. 344–420 AD) that fragments of Noah's ark had been found on the mountains of Armenia. He considered the validity of these claims to be "undecided."[75] He also evaluated the claims of "certain eminent theologians" that no rainbows appeared in the sky until after Noah's flood. These he considered "frivolous."[76] He asserted, "The words of Moses do not signify, that a bow was then formed which did not previously exist, but that a mark was engraven upon it, which should give a sign of the divine favor towards men."[77] In Calvin's view, "the celestial arch which had before existed naturally, is here consecrated into a sign and a pledge; and thus a new office is assigned to it."[78] Thus, Calvin implies that Earth's water cycle remained essentially the same after the flood as before.

Christian Scholars of the Seventeenth and Eighteenth Centuries

Book publishing emerged as a flourishing industry during the seventeenth and eighteenth centuries. This development was especially evident in Britain. One of its manifestations was the printing and widespread distribution of commentaries on the Bible. Most broadly familiar were those written by Matthew

Henry and John Wesley, commentaries that remain popular to this day. Both Henry and Wesley, in commenting on Noah's flood, treated Genesis 6–11 as a plausible account of actual events in human history *and* as an allegory for baptism and salvation.

Monolithic Interpretations

It seems worth noting, again, how little variance had developed among biblical scholars up to this time in their views on the flood. By nearly universal consensus, the correct interpretation of Genesis 6–9 was that these chapters describe real historical events. Most of these same scholars also agreed that the literal, historical message of Genesis 6–9 led to the symbolic or analogical inferences found in the New Testament teachings about redemption, salvation, and baptism. Though not all commented on these symbolic and analogical connections, none would have been likely to deny them.

Bible scholars writing prior to the seventeenth century clearly agreed on the demographic extent of Noah's flood. All concurred that the flood wiped out the entire population of ungodly people living at that time, and their animals, too. Readers see no hint of controversy or acrimony in these writings. Rarely did these scholars engage in debate or polemics about the flood account. When and if they did, their critiques were measured and courteous, not personal.

Gestation of Controversy

Not until the latter decades of the seventeenth century did debate arise over the geographical extent of the Genesis flood. Until that time, people had little or no awareness of the full extent of Earth's geography. When people, including Bible scholars, spoke or wrote or sang about "the world," "the whole world," "all over the world," they typically meant *their* world, the *known* world, or something *less than* the totality of planet Earth. The same could be said of those who wrote the Bible's 66 books. They used "world," "earth," "all the world," "all over the world," "the whole world," and "the whole earth" with reference to a particular region of Earth (see chapter 10).

The end of the seventeenth century witnessed, for the first time, the emergence of global citizens. Seafaring nations such as England, Spain, Portugal, and the Netherlands extended not only their exploration but also their commercial ventures to all Earth's oceans and seas. Geography began to develop into a mature science, followed by geology in the eighteenth century and by biology in the nineteenth century.

The following chapter examines how these developments in science prompted scholars and thinkers to integrate *new* knowledge from "the book of nature" (one of God's two books) with the collected books of Scripture, the Bible. The opportunity to integrate both sources of "truth" intrigued and challenged scholars to a degree previously unimagined. These attempts fueled the birth and development of today's controversies over how to interpret the Genesis flood.

Chapter 6

The Debate Begins

The Scientific Revolution that commenced during the Reformation era accelerated dramatically during the 1600s and 1700s. European scholars, both practicing scientists and biblical theologians, eagerly attempted detailed and specific integration of data from God's two books: the book of nature and the book of Scripture (the "books" explicitly referred to in Psalm 19 and in article 2 of the Belgic Confession, written in 1561). Their intellectual endeavors began to broaden the range of interpretive models, and these in turn sparked healthy debate. They also sowed the seeds of what later devolved into sharp conflict.

Seventeenth-Century Theologian-Scientists

The first public figures to develop a detailed integration of the biblical texts on Noah's flood with the findings of geology were Athanasius Kircher (1602–1680), Isaac Vossius (1618–1689), Edward Stillingfleet (1635–1699), Thomas Burnet (c. 1635–1715), John Woodward (1665–1728), Edmund Halley (1656–1742), and William Whiston (1667–1752). Like Isaac Newton (1642–1727), they all believed God's book of nature could be used to better understand God's book of Scripture—and, in turn, God's book of Scripture could be used to better understand God's book of nature.

Athanasius Kircher, a German Jesuit priest, served for more than 40 years on the faculty of the Jesuit College of Rome. During those years he taught mathematics, physics, and Oriental languages. He also published major works on both geology and medicine. In 1675, he published *Arca Noë* (Noah's Ark), a three-volume book in which he sought to demonstrate how geological science supported the flood account in Genesis.[1] In the first volume, *De Rebus quae*

Ante Diluvium (The Issues That Arise Before the Flood), he discussed construction of the ark, the animals taken on board, and how the animals were to be cared for. In the second volume, *De Iis, quae Ipso Diluvio e Jusque Duratione* (Of Them, the Flood and the Right Duration), he described the flood and how the ark could have survived it. In the third volume, *De Iis, quae Post Diluvium a Noëmo Gesta Sunt* (Of the Things That Came After Noah's Flood), he compared the world before the flood with the world after.

Kircher treated the details and dimensions of the ark described in Genesis as literal and accurate.[2] However, he recognized that at such a size, the ark would be inadequately spacious to accommodate all Earth's known land animal species. This awareness led him to hypothesize that only about 50 unique pairs of animals were brought on board Noah's ark. He assumed that these 50 types of animals could have produced, through hybridization, all the species of land animals alive in his day.[3]

In describing Noah's flood as "universal," Kircher meant it was sufficiently extensive to wipe out all humans and all land animals on Earth.[4] He included among the animals destroyed not only mammals and birds but also reptiles. Thus, by implication, he claimed that Noah's flood must have covered most of Earth's landmasses.

In the third volume of *Arca Noë*, Kircher contrasted the landmasses of the world before the flood with the landmasses after the flood.[5] He claimed that the floodwaters primarily came from huge underground lakes that God caused to overflow onto Earth's surface. He hypothesized that during the flood, enormous blocks of Earth's outer shell (crust) buckled up to form Earth's mountains and mountain ranges. Thus, he argued that the area and vertical relief of Earth's landmasses were greatly increased by the flood.

Dutch theologian Isaac Vossius authored more than a dozen books and treatises on science and Scripture. The best known, relevant to Noah's flood, are these three: *Dissertatio De Vera Aetate Mundi* (Dissertation on the True Age of the World), 1659; *De Septuaginta Interpretibus, eorumque Translatione et Chronologia Dissertationes* (Of the Septuagint Interpreters and Their Translation and Chronology Dissertations), 1661; and *Theologia Gentili et Physiologica* (Gentile and Physiological Theology), in 1685 and again posthumously (with slight alteration) between 1701 and 1705.

Vossius commented most extensively on Noah's flood in *De Septuaginta Interpretibus*. In this volume he demonstrated that the total known quantity of water on Earth's surface and in Earth's atmosphere and crust amount to roughly an order of magnitude less than what would be needed for liquid water to cover all Earth's landmasses for several months.[6] He considered as utterly foolish—and counter to God's character—any proposal that God miraculously created all the extra water needed to cover Earth's surface and just as miraculously removed it all, without leaving a single trace of either miracle.

Vossius proposed, instead, that the flood need not have been globally extensive if humanity inhabited just one or a few contiguous regions of Earth at Noah's time. He emphasized that to punish or destroy parts of Earth untouched by humans and their animals would be counter to God's revealed character. He pointed out that the flood would still be "worldwide" if it covered the then-known world, the part of Earth inhabited by human beings and their animals. He observed that the Bible frequently refers to local events as impacting the "world" or "all the world" (see chapter 10).

Vossius went on to cite biblical texts and what was known (at the time of his research) about the history of human migration, colonization, and population to argue that the geographical extent of humans and their animals at the time of Noah would have been limited to the Middle East. On this basis, he concluded that Noah's flood covered most of the Middle East but probably only little more beyond.

Several of Vossius's contemporaries also maintained that the Genesis flood must have been regional, rather than global.[7] Edward Stillingfleet, Anglican Bishop of Worcester and a frequent speaker in the House of Lords, was among them. He, too, held that Noah's flood was regional, not global, and that such an interpretation of Genesis 6–9 was fully consistent with biblical accuracy and orthodoxy.

In 1666, Stillingfleet completed the third edition of his book *Origines Sacrae* (Sacred Origins) in which he commented extensively on the flood. He argued that the Bible taught a "universal deluge,"[8] but only with respect to mankind. "Whether universal as to the earth and animals," he saw "no necessity of asserting either."[9] He then calculated that all the water in Earth's clouds would deliver, at most, only 18 inches of precipitation—if spread out over Earth's entire surface, hardly enough to submerge all Earth's landmasses.[10] He calculated that provision of new water to overflow all Earth's surface would require the miraculous creation and later removal of 15 times the quantity of water contained in all Earth's oceans, seas, lakes, and rivers.[11] Such an appeal, stated Stillingfleet, is in no way mandated by Scripture. In fact, he argued, it could be ruled out, biblically.

Noah's ark, Stillingfleet pointed out, *if* built according to the dimensions set forth in Genesis 6 would be much too small to accommodate pairs of all

Earth's land animals and the provisions needed to feed them for a year and 17 days (the total time Noah, his family, and the animals were on the ark, according to Genesis).[12] However, he agreed with Vossius that such an ark would be adequate for eight people, their animals, and their food—provided that the area covered by water was limited to the environs of the Middle East.[13]

Stillingfleet wrote that, "it is most certain according to Scriptures" that "all mankind (those on the Ark excepted) were destroyed by it [Noah's flood]."[14] He went on to conclude, "So then, it is evident that the Flood was *universal* as to mankind, but from thence follows no necessity at all of asserting the universality of it as to the *Globe* of the *earth*, unless it be sufficiently proved that the whole earth was peopled before the *Flood* (emphasis in the original)."[15] Such proof, according to Stillingfleet, was lacking, both physically and biblically. He concluded, "I see no reason to extend the destruction of these beyond that *compass* and *space* of earth where men inhabited" (emphasis in the original).[16]

Additionally, Stillingfleet asserted that the animals on board Noah's ark could not possibly have come from all Earth's landmasses. He explained that "some kind of *Animals* cannot live out of that particular *Clime* wherein they are" (emphasis in the original).[17] Noah's ark, Stillingfleet said, could not have offered all the "climes" needed for the various creatures' survival for more than a year. Nor would it have been possible for most animals to remain healthy enough to survive their long-distance migrations back to their original climate zones.[18]

Thomas Burnet, royal chaplain to King William III, also commented extensively on Noah's flood in his best-known book, *Telluris Theoria Sacra* (Sacred Theory of the Earth) published in two volumes between 1681 and 1689. Although a few Christian scholars had earlier expressed belief that Noah's flood had covered the whole planet, Burnet was the first well-known Christian scholar to do so. More importantly, he was the first Christian scholar to propose a model explaining *how* Noah's flood could have been globally extensive.

After making the claim that Noah's flood "overspread the face of the whole Earth, from Pole to Pole, and from East to West,"[19] Burnet recognized by his own calculation that all the water on Earth—even with an additional 40 days and nights of rain—would be insufficient to account for global coverage.[20] Therefore, he proposed that before the flood, Earth's surface "was smooth, regular and uniform, without Mountains, and without a Sea."[21]

He further speculated that Earth was hollow, with nearly all its water trapped inside an outer shell until the time of Noah's flood.[22] He hypothesized that Earth's mountains formed during the flood from the broken fragments of the outer shell, further suggesting that the receding of the waters, described in Genesis 8, referred to the return of most of the floodwaters back into the hollow core of Earth.[23]

Burnet's interpretation gained momentum when taken up by John Woodward. A naturalist, geologist, medical doctor, professor of medicine and physics, and member of the Royal Society (the world's first national academy of sciences), Woodward was England's foremost expert on fossils. In 1695, he published a book on Earth's natural history, almost wholly devoted to a detailed discussion of Noah's flood.[24] In Part VI of this book, he commented on Burnet's work and later went on to write far more about Noah's flood than any scholar who preceded him.

Woodward explicitly and repeatedly claimed that Noah's flood had been a global event, that "did actually cover all Parts of the Earth,"[25] that "covered the Terrestrial Globe,"[26] and "was *Universal*, and laid the whole Earth under Water: covering all, even the highest, Mountains, quite around the Globe" (emphasis in the original).[27] Woodward was convinced that all fossils were the remains of organisms that perished in Noah's flood.[28] In his words, "[By] a deliberate and careful Examination of all Circumstances of these Marine Bodies [fossils], I was abundantly convinced that they could not have come into those Circumstances by any other means than such a Dissolution of the Earth."[29] He asserted that the flood wiped out both land and sea animals.[30] He also saw the flood as responsible for all Earth's metal, mineral, and fossil fuel deposits.[31]

Woodward hypothesized, as Burnet had, that a huge subterranean abyss filled with water broke through the earth's crust. In his scenario, this break through did more than rip apart the crust.[32] He wrote that because of it, "the Terrestrial Globe was entirely dissolved."[33] He went on to propose that the cause of the breakthrough was "extraordinary Discharges of the Subterranean Heat"[34] that had been trapped in Earth's deep interior.[35] The postflood outcome, he suggested, was "that the *whole terraqueous Globe* was, by this means, put into the Condition that we now behold" (emphasis in the original).[36]

According to Woodward, the densest parts of Earth's "dissolved" crust settled out as the floodwaters receded and became progressively less dense through time.[37] In this way, the density of layers comprising Earth's new crust increased with depth. Based on this hypothetical model, he predicted that the densest fossils would be found in Earth's deepest layers and the least dense fossils in the top layers.[38]

Noah's flood, Woodward declared, accomplished not one but two primary purposes. He agreed with Christian scholars of the previous 15 centuries that God's purpose for the flood was to judge Noah's contemporaries for their rampant evil and, thus, to save humanity from becoming completely corrupt. However, he claimed that his predecessors had missed an equally or even more important purpose: radical transformation of Earth's surface. He wrote, "Altho' one Intention of the Deluge was to inflict a deserved Punishment upon that Race of Men, yet it was not solely levelled against Mankind, but principally against the Earth that then was; with design to destroy and alter that Constitution of it."[39]

Woodward reasoned that humans living before Noah's flood had degraded to pervasive, unspeakable evil because their life was too easy.[40] He surmised that life had been "Paradise before, but was to be none after the Deluge."[41] While acknowledging that antediluvian conditions would have been less pleasant than those in the Garden of Eden, he proposed that before the flood "in all parts of it [Earth] was yet generally much more fertile than ours is,"[42] and Earth's "Soil was more luxuriant, and teemed forth its Productions in far greater plenty and abundance than the present Earth does."[43] Thus, Woodward assumed, "That Earth, requiring little or no Care or Culture; but yielding its increase freely, and without considerable Labour or Toil, or assistance of Humane Industry" had led to ease.[44] He concluded that the whole of humanity before the flood simply reaped the always readily available bounty of food. In his view, "Their Agriculture was nothing near so laborious, and troublesome, nor did it take up so much time as ours doth."[45] He saw this ease of living as applying to all Earth's landmasses. Therefore, he believed, humanity had become globally dispersed before Noah's flood.

In its dramatically transformed surface condition, Earth would require hard work of those who sought to eke out a living, Woodward reasoned. He saw such universal hard labor as necessary to prevent a return of the horrifying evil of Noah's contemporaries. After all, with "Earth requiring little or no Tillage, there was little occasion for Labour; so that almost his [humanity's] whole time lay upon his hands, and gave him leisure to contrive, and full swing to pursue his Follies; by which means he was laid open to all manner of Pravity, Corruption, and Enormity."[46] After the flood, this "general Employ and Expense of their time . . . would shorten the Opportunities of Vice."[47]

Woodward argued that if Noah's flood had not radically transformed the entirety of Earth's surface, humanity would have suffered radical judgment more than once. From his perspective, unless Earth had been altered in this way "there would have been a new necessity to punish and reclaim the World; to depopulate the Earth and reduce it again to a vast Solitude, as constantly as there succeeded a new Age and Race of Men."[48]

Clearly, Woodward's model for Noah's flood is rooted in his assumption that the physics, geology, and chemistry of Earth's surface and crust were drastically and permanently degraded by Noah's flood. However, the Bible contradicts this assumption in at least two ways. First, Genesis 3 states that the ground was cursed from the moment Adam sinned and that this curse would result in substantially more work and more pain for humanity. Therefore, the preflood era would not have been the life of ease, plenty, and leisure Woodward envisioned. Second, passages such as Ecclesiastes 1:4–10, 3:11–15, Jeremiah 33:25, and Romans 8:18–23 affirm that the physical constants governing Earth have not changed (see chapter 18).

Edmund Halley, another distinguished member of the Royal Society and Britain's Astronomer Royal, shared Woodward's belief that Noah's flood was a global event. However, he offered a different explanation of the floodwaters' source (and purpose). Best known for determining the orbits of comets and predicting the return of the comet named after him, Halley proposed an astronomical model for delivery of the flood's water. In two papers presented at Royal Society meetings in 1694,[49] Halley shared his calculations showing that 40 days and nights of torrential rain, plus waters rising up from underground aquifers would be insufficient to cover all Earth's continents and islands.[50] Halley hypothesized that the additional water required for global coverage came from the "Choc [tail and corona] of a Comet, or other transient Body."[51]

According to Halley's explanatory model, this Choc, or other body, did much more than provide water. It also served as "an Expedient to change instantly the Poles and Diurnal Rotation of the Globe."[52] Such a change in the tilt of Earth's rotation axis and in Earth's rotation period, Halley claimed, would have caused the waters in Earth's oceans to slosh over all Earth's landmasses, "raising up Mountains where none were before."[53] This action would supposedly account for the existence of marine fossils on Earth's continents.

Beyond its attempt to account for the water's source and the presence of seashells on mountaintops, Halley's model offered no interpretive insights. He openly acknowledged "difficulties" for his interpretation and explanation of Noah's flood.[54] The construction and size of Noah's ark, as well as the need to preserve representatives of all Earth's land animals on the ark posed challenges for which he offered no plausible answers.

William Whiston, successor to Isaac Newton as Lucasian Professor of Mathematics at Cambridge University, like Newton, considered the Bible, including the Genesis accounts of creation and the flood, as literally and historically true. Whiston, aware of Halley's presentation before the Royal Society, incorporated some of Halley's ideas in a series of books, *New Theory of the Earth*, published in 1696. In this series, Whiston asserted that the tilt of Earth's rotation axis prior to Noah's flood had been more extreme, by far, than its later (and still current) 23.5 degrees. This greater tilt, he argued, would have rendered most of Earth's surface uninhabitable by humans and unsuitable for growing fruits and vegetables.[55] He speculated that Noah's ark had come to rest on the Caucasus Mountains, not on the mountains of Armenia.[56] He posited that all parts of the world inhabited by humans were covered by waters from the comet's atmosphere when Earth passed through the comet's tail.[57] He went on to suggest that the comet's gravitational force would have been sufficient to draw subterranean waters from beneath Earth's crust onto the surface.[58]

Whiston agreed with his colleague Burnet that the floodwaters reached "all parts of the Earth, and destroying all the Land-animals on the intire Surface thereof: those only excepted which were with Noah in the Ark."[59] He additionally claimed that the floodwaters were calm in the vicinity of the ark,[60] while elsewhere huge, violent waves sloshed over the plains, hills, and mountains. His writings fail to clarify whether he believed the waters ran intermittently over various regions or remained over Earth's entire surface for several months.

The World of Burnet, Woodward, Halley, and Whiston

Thomas Burnet, John Woodward, Edmund Halley, and William Whiston were the first widely known and respected scholars to declare the Genesis flood a globally extensive event. All four were Englishmen, personally acquainted with Sir Isaac Newton, and known to one another. In their time, at the seventeenth century's end, Britain ruled the waves and became the first globally expansive empire. The positions and status of these men within British academia and British politics made them global citizens. Halley made astronomical observations in both the Northern and Southern Hemispheres.

Burnet, Woodward, Halley, and Whiston were also mavericks, some might say "heretics," within Christian academia. Their nonconformist beliefs became evident in their speculations, especially about Earth's geological structure and history. Nevertheless, they freely admitted that they were speculating. Their writings on Noah's flood express none of the dogmatic certitude so common among global flood proponents of the past century, right up to today.

Scholars in the latter half of the seventeenth century witnessed—or participated in—the birth of physics as a serious scientific discipline and the founding of the Royal Society. The eighteenth century saw the birth of geology as a serious scientific discipline. Tolerance of nonconformist ideas and experimental hypotheses greatly increased. A movement known as the Enlightenment swept through Europe during the eighteenth century. Its central emphases included individual liberty, religious tolerance, and opposition to the orthodoxy (or dogma) of both the Roman Catholic and Protestant state churches. As this movement grew, scholars at universities and seminaries as well as other public leaders found it acceptable, even valiant, to deny the authority and inerrancy of the Bible.

The Enlightenment opened wide the door to heterodoxy in Christian seminaries and to deism and atheism in universities. It paved the way for new controversies and rivalries to develop, many of which later generated sharp and/or hostile divisions both inside and outside the Christian community.

Eighteenth Century Bible Commentaries

During this same era, as books and education became more accessible to more people, commentaries on the Bible, particularly those written by Matthew Henry and John Wesley, became widely read and studied. These commentaries, which included none of the scientific and theological speculations of Burnet, Woodward, Halley, and Whiston, strongly influenced what Christians thought about Noah's flood.

Matthew Henry (1662–1714), a British Presbyterian minister, produced the first complete commentary on the Bible. Originally published as a six-volume set in 1706–1710, it *still* ranks as the most popular and influential commentary on every verse of the Bible.

Henry commented extensively on Noah's flood in volume 1, which covered Genesis. In his literal, historical interpretation of Genesis 6–9,[61] he wrote that apart from Noah and his family, the entire human race had become bent on "nothing but murders, rapes, and rapine."[62] He described Noah as a preacher to his generation, warning of judgment and calling people to repentance, receiving only ridicule and laughter in response.[63] Like Luther, Henry interpreted the depiction of the ark in Genesis 7 as accurate in all its dimensions and details.[64] He wrote, "Perhaps there needed no new creation of waters"[65] and "nothing is more needful or useful than waters, both the springs of the earth and the showers of heaven."[66] With these words he indicated that the rain from above and water from underground aquifers would have been sufficient under God's control to destroy all humanity and their animals, other than those on the ark.

Like most of the early church fathers, Henry saw analogies between the historical event of the flood and the message of salvation through Jesus Christ.[67]

The height and extent of Noah's flood, Henry asserted, were such that neither humans nor their animals outside the ark could escape.[68] All perished, said Henry, just as no one escapes the judgment of God, except in Christ.

John Wesley (1703–1791), English theologian and leader of a revival movement called Methodism within the Church of England, also wrote a commentary on Noah's flood. He, too, interpreted Genesis 6–9 as a literal, historical account. In his relatively brief work, he noted that the evil expressed by Noah's generation was unprecedented. He described the level of sin as "most gross and heinous, and provoking: committed daringly with a defiance of heaven."[69] He quoted from Matthew Henry: "There was nothing but murders, rapes, and rapine."[70]

Wesley sided with Henry in declaring the ark's dimensions both literal and accurate.[71] He expressed no doubt as to the existence of hills and mountains before Noah's flood.[72] Wesley expressed his belief that the animals destroyed in the flood were "the servants of man, and part of his possession."[73] According to Wesley, God found it necessary to destroy them because they had been "abused to be servants of sin."[74] Wesley accepted the statement that the ark came to rest on the mountains of Ararat, now known as Armenia.[75]

Birth of Geological Science

A few decades before Henry's and Wesley's Bible commentaries became available, the foundations of geological science were being laid. British scholar Robert Hooke (1635–1703) and Danish scientist Niels Steensen (1638–1686), better known as Nicolaus Steno, were the first known scholars to make a scientific case for fossils as the remains of once-living creatures turned to stone.[76] Being familiar with the biblical story, Hooke observed that the duration of Noah's flood was much too brief to explain the abundance of marine fossils and the thickness of fossil beds discovered on dry land.[77]

Steno was the first scientist known to theorize that the strata in Earth's crust were laid down at different times, with the older layers typically on the bottom and the younger, on top. He also noted observable evidence for overturning.[78] Today's geologists refer to this ordering of Earth's strata as Steno's law of superposition.

The founding of the Geological Society of London in 1807 marks the advent of geology as a mature science. By then, geological research had yielded substantial evidence that Earth's crustal features couldn't have resulted from a single event in Earth's recent history. Geologists were discovering major geological features on all Earth's continents that had formed at different times throughout Earth's history and in different ways. At the same time, there was growing recognition that humanity couldn't have inhabited all Earth's continents and islands at the widely assumed time of Noah.

Much of the credit for the birth of geology as an established scientific discipline goes to Scottish geologist James Hutton (1726–1797). His book *Theory of the Earth* laid the foundations of geological principles that still stand firm to this day.[79] These principles challenged claims that virtually all Earth's geological and geographical features formed within a single year less than 10,000 years ago. They also made clear that the enormous abundance and diversity of Earth's past life, as revealed in the fossil record, simply can't be squeezed into a period as brief as one year.

In a lecture delivered at Oxford University in 1819, English theologian, paleontologist, and geologist William Buckland (1784–1856) referred to Noah's flood as a global deluge for which there was much geological evidence.[80] He cited the observations and written conclusions of the famous French paleontologist Georges Cuvier (1769–1832).[81] However, in the same lecture, Buckland strongly rejected the idea that Noah's flood was responsible for the vast majority of Earth's past and present geological features.[82] He argued that only a small fraction of Earth's crustal strata could possibly be attributed to the flood. In his words, "It seems therefore impossible to ascribe the formation of these strata to a period so short as the single year occupied by the Mosaic deluge."[83] Like Cuvier, Buckland believed that God had intervened throughout Earth's long history before the human era—via multiple, time-separated geologic events—by adding, subtracting, and replacing various life-forms. Cuvier, Buckland, and others did not, as some Christians have suggested, reject the occurrence of catastrophic geological events.

Several years later, in a Bridgewater Treatise on the power, wisdom, and goodness of God as manifested in creation, Buckland presented additional geological and paleontological evidence to show that *none* of the strata in Earth's crust could be attributed to Noah's flood.[84] His conclusion: the notion that Noah's flood covered the entire globe is not and cannot be supported by the geological evidence.[85]

In the early nineteenth century, Scottish geologist Charles Lyell (1797–1875) reinforced and extended Hutton's geological principles. Lyell's book *Principles of Geology* amassed a wealth of geological data to demonstrate that Earth's geology was shaped predominantly, though not exclusively, by multiple slow, gradual processes.[86] In 1837, philosopher, mathematician, physicist, and theologian William Whewell (1794–1866) labeled this view "uniformitarianism."[87] Lyell's research findings, like Buckland's, dismissed the possibility that a single, recent flood could be responsible for all Earth's surface structures.

A year following the publication of *Principles of Geology*, another well-known and respected theologian-geologist, Adam Sedgwick (1785–1873), changed his formerly held views on Noah's flood, no longer attributing most of Earth's geological features to the effects of this one event. In his final presidential lecture to the Geological Society of London he spoke about his change of position:

> Our errors were, however, natural, and after the same kind which led many excellent observers of a former century to refer all the secondary formations of geology to the Noachian deluge. Having been myself a believer, and, to the best of my power, a propagator of what I now regard as philosophic heresy, and having more than once been quoted for opinions I do not now maintain, I think it right, as one of my last acts before I quit this Chair, thus publicly to read my recantation. [88]

"Scriptural Geologists"

Sedgwick's recantation, supported and lauded by Buckland and Lyell, drew strong reaction from a group of Anglican and Presbyterian theologians. Two of them, George Bugg (1769–1851) and George Young (1777–1848), responded by writing books that gave rise to the term "scriptural geology."[89] Of the 25 individuals who comprised this group,[90] only two (John Murray and George Young) had done any geological research. The remainder had no formal training in geology, though two (George Fairholme and Granville Penn) had done considerable reading in geology.

Theologians who followed in their path and identified themselves as scriptural geologists committed to upholding the following:

1. biblical inerrancy in all matters that the Bible addresses,
2. a literal, historical interpretation of Genesis 1–11,
3. rejection of uniformitarian geological principles in favor of catastrophism,
4. rejection of multiple geological catastrophes in favor of one major catastrophe,
5. a 6,000–10,000-year age for the universe and Earth, and

6. Noah's flood as globally extensive and responsible for nearly all Earth's geological features.

In the eyes of scriptural geologists, to give way on any of these tenets was tantamount to impeaching the Bible as God's Word to humanity. As one of their adherents David Lord wrote in his book *Geognosy*, "The contradiction which the modern theory of geology presents to the record of creation by Moses, naturally leads those who assent to it, to regard that record as erroneous and prepares the way for a distrust and rejection of the whole Bible."[91]

Buckland, Sedgwick, and their many allies in the Geological Society of London firmly rejected their implied new label as "antiscriptural geologists." They affirmed that, like the so-called "scriptural geologists," they upheld the inspiration and inerrancy of the Bible in all that it addresses. They affirmed their belief that Genesis 1–11 is an accurate historical record. The only issue they disagreed on was what constituted a "literal, historical" interpretation of Genesis 1–11 (see chapter 12). Buckland saw the need to make this public statement:

> If the suggestions I shall venture to propose require some modification of the most commonly received and popular interpretation of the Mosaic narrative, this admission neither involves any impeachment of the authenticity of the text, nor of the judgment of those who have formerly interpreted it otherwise, in the absence of information as to facts which have but recently been brought to light. [92]

The Near Demise of Scriptural Geology

Between 1820 and 1850, the scriptural geologists group published more than 50 articles and books. Then, from 1850 to 1860, the movement rapidly declined and all but disappeared. Among the principal reasons for this decline are these four:

1. lack of agreement on where to place Noah's flood in the geological strata,
2. George Bugg's change of position from rejecting to accepting that *some* of the fossil record may have preceded Adam's creation and perhaps even the six days of creation,[93]
3. introduction of the "appearance of age" hypothesis by Philip Gosse

(1857) to explain away geological dates in the millions of years for Earth's strata and fossils,[94] and

4. publication of Charles Darwin's *On the Origin of Species* (1859),[95] which sensationally diverted attention away from Noah's flood and toward questions about life's origin and history.

For the next half century, the scriptural geology movement remained essentially comatose. Then, a Seventh-day Adventist antievolutionist brought it back to life, reigniting the controversy over Noah's flood. The next chapter continues the story of how the Genesis flood became a heated controversy—and, for many people today, a stumbling block on the path to faith.

Chapter 7

Controversy Reignited

In 2015, University of Chicago professor Jerry Coyne published a book, *Faith v. Fact*, celebrating Charles Darwin's work as "the deadliest blow ever struck by science against faith."[1] Coyne hailed *On the Origin of Species* as "the greatest scripture-killer ever penned."[2] According to Coyne, Darwin had demolished "an entire series of biblical claims by demonstrating that purely naturalistic processes—evolution and natural selection—could explain patterns in nature previously explainable only by invoking a Great Designer."[3]

Most Christian scholars of the latter nineteenth century considered *On the Origin of Species* no major threat to their confidence in the truth of Scripture. With rare exceptions, British creationists of the Victorian era had little trouble assimilating the findings of geologists and paleontologists James Hutton, William Buckland, Charles Lyell, and Adam Sedgwick into their understanding of the Bible's creation texts. Likewise, in America, respected Christian theologians including Charles Hodge, James Orr, and Luther Townsend, as well as geologists Louis Agassiz and George Frederick Wright, easily incorporated the evidence of life's progression observed in the fossil record into their interpretations of biblical creation texts. It seemed reasonable to view God's creative work as occurring progressively over time. In no case did any of these creationists see a need to invoke Noah's flood as the explanation for Earth's fossil record.

Descent of Man

Charles Darwin's 1871 publication, *The Descent of Man*,[4] met with a more divisive response from these and other theologians. In this book, Darwin had dropped any pretense of the Creator's involvement in shaping life. He expressed certainty that humans did not descend from a man and a woman specially created by God, creatures who were no different from today's humans. Rather, he asserted, "man is descended from some lowly organized form,"[5] more specifically, "from a hairy, quadruped, furnished with a tail and pointed ears, probably arboreal in its habits."[6]

Darwin had thrown the proverbial gauntlet. While some scholars, including Thomas Huxley, cheered a challenge to the authority of the church, other scholars, Christian and otherwise, recognized Darwin's hypothesis as an overly bold declaration still open to question and ongoing assessment. The debate over "truth" that had previously unfolded between Galileo Galilei and the Roman Catholic Church now came to Britain.

It was fueled by highly publicized "debates" that began as open discussions, such as the one between Darwin's friend Huxley and Samuel Wilberforce, Bishop of Oxford. Periodicals of the day highlighted and dramatized animated interactions between scholars of science and scholars of Scripture. The Church of England, which had been home to the scientific establishment, began to divide between those who welcomed and embraced the "advance of science" and those who dismissed or feared it. This division grew stronger over time, forming two major branches of Christianity, the proverbial "liberal" and "conservative" branches, loosely defined by their approach to God's two books, nature and Scripture.

Darwin's words drew a particularly emphatic response from an emerging ultraconservative Christian sect known as the Seventh-day Adventists. In 1876, one of their leaders, Reverend P. R. Russel, published an article in *The Advent Review and Sabbath Herald*, expressing his and his peers' indignation: "The Scripture record places a crown of honor and dominion on the brow of our common humanity. . . . Darwinism casts us all down from this elevated platform, and herds us all with four-footed beasts and creeping things. It tears the crown from our heads; it treats us as bastards and not sons and reveals the degrading fact that man in his best estate—even Mr. Darwin—is but a civilized, dressed up, educated monkey, who has lost his tail."[7]

Early Adventist Theology

One of the most significant tenets that distinguished Seventh-day Adventists from other Protestant, Catholic, and Orthodox Christians was belief that to be a true follower of Jesus Christ one must keep "the Sabbath," which Adventism identified as Saturday, not Sunday, or any other day of the week, for that matter. Ellen White (1827–1915), cofounder, prolific writer, and bold speaker within this new group, gained prominence when she wrote, authoritatively, that

Exodus 20:11 confirms Adventist doctrine.[8] White said she received a vision from God, revealing to her "that the first week [of the world's history], in which God performed the work of creation in six days and rested on the seventh day, was just like every other week."[9]

Considered by many Adventists as a prophet, White taught that Noah's flood had been a global, catastrophic event responsible for completely reshaping the planet. She wrote, "The entire surface of the earth was changed at the Flood. . . . In many places hills and mountains had disappeared, leaving no trace where they once stood; and plains had given place to mountain ranges."[10] She also attributed Earth's immense coal and oil deposits to the flood's transformation of the forests that had existed prior to the deluge.[11] She dismissed the findings of contemporary geologists: "Apart from Bible history, geology can prove nothing."[12] Although she rejected much of what scientists were reporting, she did endorse the doctrine of God's trustworthy revelation in two books, nature and Scripture. She declared, "All truth, whether in nature or in revelation is consistent with itself in all its manifestations."[13]

Through the first half of the twentieth century, the Seventh-day Adventist denomination held firmly to White's teaching on the Sabbath—as well as on other topics, including God's recent creation of the world and Noah's flood. Many Seventh-day Adventists still do. However, today there is more openness among Adventist theologians and church leaders. Few condemn other Christians who celebrate the "Sabbath" on Sundays and view God's creation "week" as something other than a 168-hour work period sometime between 6,000 and 10,000 years ago.

Many Seventh-day Adventist theologians and professors at Andrews University, Loma Linda University, and the University of the Southern Caribbean endorse belief in "day-age creation" (the idea that God created over six long epochs followed by an ongoing period of cessation from physical creation) and accept the view that Noah's flood, while deadly to all ungodly humans (2 Peter 2:5), is not responsible for all Earth's geological features. Others, however, still hold to White's teachings on creation and Noah's flood.

George McCready Price

The most successful promoter of White's views on Noah's flood was George McCready Price (1870–1963). After the death of his father in 1882, his mother raised him in the Seventh-day Adventist church. In his early twenties, Price attended Battle Creek College (now Andrews University) the Adventist denomination's flagship institute of higher learning. From there, he enrolled in a one-year teacher preparation course at the Normal School of New Brunswick (now the University of New Brunswick). There he took some basic courses in science, including an introduction to geology.

Price, an ardent opponent of natural-process evolution, sought to undercut Charles Darwin's ideas about how Earth's life came to be. Like many Christians of his time, he assumed that an ancient Earth served as a key prop in support of Darwinism. Based on this assumption, he considered that the optimal weapon against Darwinian claims was the Seventh-day Adventist doctrine of a relatively recent creation.

Price became convinced that geology would be the best scientific tool for proving that Earth is much too young to provide any basis for belief that the origin and history of all life on Earth came about through strictly natural means. In Price's brief exposure to geology, he figured that professional geologists must have misinterpreted their data. He reasoned that they must be wrong about the fossils having been laid down at different times throughout Earth's long history.

In 1906, Price published *Illogical Geology: The Weakest Point in the Evolution Theory*.[14] In this book, he described "the phenomenal weakness of the geological argument" for "the long devious processes connected with Darwinism and biology."[15] In the introduction he wrote, "I am willing to give a thousand dollars to anyone who will, in the face of the facts here presented, show me how to prove that one kind of fossil is older than another."[16] When no one claimed his extravagant prize, he viewed his position as unassailable, failing to realize that no research geologist could agree with the "facts here presented." Undaunted, Price went on to publish *The Fundamentals of Geology* in 1913 and *The New Geology* in 1923.[17] This book, a greatly expanded version of *Illogical Geology*, asserted that Noah's flood was responsible for virtually all Earth's fossils.

Resurgence of "Flood Geology"

The impact of Price's ideas spread far beyond the Seventh-day Adventist schools and congregations. *The Fundamentals of Geology*, though soundly rejected by Christian geologists, was highly praised by several early twentieth-century Protestant Christian apologists, including Harry Rimmer, the best-known Christian creationist and apologist of the first half of the twentieth century. Despite disagreeing with Price's views that Noah's flood was global in extent and that Earth is younger than 10,000 years old, Rimmer declared *The Fundamentals of Geology* "a masterpiece of REAL science."[18]

During the infamous Scopes Trial (1925) over the legality of teaching human evolution, prosecuting attorney William Jennings Bryan, three-time presidential candidate and former secretary of state, made extensive use of Price's work. Consequently, Price's argument for Noah's flood as a globally extensive event responsible for nearly all Earth's fossils and geological features catapulted to worldwide notoriety.

Riding this wave of public attention and interest, *The Genesis Flood*,[19] a book published in 1961 and coauthored by theologian John C. Whitcomb and engineering professor Henry M. Morris II, eventually gained a far wider audience than Price's work. After all, *The Genesis Flood* came from "credentialed experts" who represented and appealed to a broad spectrum of Protestant denominations. Whitcomb had taken undergraduate geology and paleontology courses at Princeton University, and Morris had professional expertise in hydraulics engineering.

The Genesis Flood was a densely written tome filled with footnotes, endnotes, and scientific jargon. For lay readers, it had an air of scientific respectability. It linked its geological claims with specific biblical texts. It purported to provide a strong defense of biblical inerrancy. Any scientific critique of its claims was blunted by the authors' intimation that mainstream science had become a godless cabal committed to deceiving the public into belief that naturalistic processes, alone, could explain the origin and history of life—as well as all Earth's geological and chemical features.

With the founding of the Institute for Creation Research (ICR) by Henry Morris in 1972, *The Genesis Flood* gained popularity among Christians worldwide. Over 300,000 copies of the book's 48 printings have been sold, and it has been translated and distributed in multiple languages.

The book reads like an updated and expanded version of Price's *The Fundamentals of Geology*, featuring multiple new "insights." It claims that before Adam and Eve sinned in the Garden of Eden, animals around the world experienced no death, no disease, and no suffering. In fact, it says humans and dinosaurs overlapped in their time on Earth. It further asserts that the second law of thermodynamics (law of increasing entropy, or decay) did not begin to operate until *after* Adam and Eve sinned.

Many years later, at a lecture given in the early 1990s, Morris, to his credit, admitted that his claim about the second law of thermodynamics having come into effect only after Adam sinned was demonstrably untenable.[20] He stated as his new position that God accelerated the second law of thermodynamics when Adam sinned. However, he then proceeded to dismiss allegations that his revised position is no more scientifically tenable than the original.

An Ironic Development

Science historian Ronald Numbers noted that because of *The Genesis Flood*, "geological issues push biological ones to the periphery, and the Noachian deluge tends to eclipse the Adamic creation in importance. 'The Genesis Flood is the real crux of the conflict between evolutionist and creationist cosmologies.'"[21]

Focus on Noah's flood became more intense in 2002 when geophysicist John Baumgardner joined the ICR staff. Baumgardner brought with him his attempt to demonstrate *how* Noah's flood could explain nearly all Earth's geological features.[22] He hypothesized (without the backing of physical evidence) that the slip-sliding of Earth's tectonic plates moved millions of times faster during Noah's flood than they move today. He and his colleagues proposed that radiometric decay rates—if increased by a factor of about a billion times during a large portion of the Genesis flood—could explain this accelerated plate movement.[23]

"Flood Geology" Becomes Entrenched

The supercontinent cycle described by mainstream geologists is a quasiperiodic joining together of all Earth's continents and islands into one supercontinent and the subsequent splitting apart of this supercontinent into multiple spacially separated continents and islands. Evidence shows that five such cycles have occurred in Earth's history.[24] While so-called flood geologists agree with mainstream geologists that such a cycle has operated in Earth's history, they insist that at least one, if not two, supercontinent cycles were compressed into the single year of Noah's flood (see chapter 13).

Flood geologists today assert that Noah's flood not only brought about the splitting apart of Earth's currently existing seven continents, but also the formation of virtually all Earth's islands, lakes, mountain ranges, canyons, volcanoes, sedimentary plains, plateaus, and lava fields. They claim, further, that all this geological activity occurred during the year Noah and his family spent on board the ark.[25]

Rather than try to compress all Earth's ice ages (periods of extensive ice coverage) into a single year, flood geologists vehemently deny that Earth ever experienced an ice age cycle. Meanwhile, geologists and climatologists have analyzed ice core records, observed cyclical changes in Earth's rotation axis tilt and orbital eccentricity, and discovered evidence of past asteroid and comet impacts. All such research points to the conclusion that more than 50 ice ages and interglacials (warmer periods between ice ages) have occurred during the past 2.58 million years, not to mention an unknown number in earlier eras.[26]

Their research shows that these ice ages lasted several tens of thousands of years, with the interglacials lasting several thousands of years. Flood geologists contend that Earth has experienced only a single ice age that lasted only some hundreds of years.[27]

Flood geologists describe this one ice age as a brief, catastrophic climate change episode that occurred immediately after Noah's flood.[28] They propose that as the waters of Noah's flood receded, volcanic eruptions all over the world filled Earth's atmosphere with sunlight-blocking aerosols, and as these volcanic eruptions continued for several hundred years, thick ice sheets formed over Earth's high northern and southern latitudes.[29] Once the volcanic aerosols cleared from Earth's atmosphere, they say, Earth quickly returned to the climate conditions that existed before Noah's flood.

Response to Biblical Inerrancy Statements

Flood geologists insist that the affirmations and denials produced by the International Council on Biblical Inerrancy during the 1970s and 1980s (see chapter 1) are less than sufficiently stringent. Thus, they have published additional affirmations and denials designed to make their interpretation of Genesis 6–9 a requirement for adherence to the doctrine of biblical inerrancy. These augmented statements relevant to Noah's flood are as follows:[30]

> IV. **We affirm** that no apparent, perceived, or claimed evidence in any field, including history, archaeology, and science, can be considered valid if it contradicts the scriptural record. We also affirm that the evidence from such fields of inquiry is always subject to interpretation by fallible people who do not possess all information.
>
> **We deny** that scientific "evidence" used to "prove" millions of years is objective fact and not heavily influenced by naturalistic presuppositions.
>
> XI. **We affirm** that the great Flood described in Genesis 6–9 was an actual historic event, worldwide (global) in its extent and extremely catastrophic in its effect. As such, it produced most (but not all) of the geological record of thousands of meters of strata and fossils that we see on the earth's surface today.
>
> **We deny** that Noah's Flood was limited to a localized region (e.g. the Mesopotamian valley of the Tigris and Euphrates Rivers). We also

> deny that the Flood was so peaceful that it left no abiding geological evidence. We further deny that the thousands of meters of sedimentary rock formations with their fossilized remains were largely produced after or before the Flood or even before Adam.

Scholarly Response to Flood Geology

The twenty-first century scientific community denies flood geologists' claims, but not based on anti-Christian or antibiblical bias. Rather, the research community sees no credible scientific evidence to support flood geologists' hypotheses. At the same time, the scientific community sees flood geologists' willingness to contradict firmly established findings—including biblically affirmed truths, such as the constancy of physical laws. Flood geology clashes with the fossil record and established dating methods and falsely accuses geologists of being strictly "uniformitarian" in their analysis of data. In reality, geologists, paleontologists, and biologists argue for a combination of *both* enduring processes whereby gradual change occurs, and multiple catastrophic events resulting in dramatic and rapid changes.

The scientific community has found no evidence to support flood geologists' belief that a singular catastrophic event within the last ten thousand years is responsible for virtually all Earth's geological features and history. Any indication that either the laws or constants of physics have been significantly altered during Earth's history would be unmistakably obvious. They see no such indication.[31]

Flood geologists' proposed additions to the affirmations and denials published by the International Council on Biblical Inerrancy have been rejected by the majority of Christian theologians. Nevertheless, the popularly reported controversy over their claims has given rise to a broad range of questions and speculations about what the Bible actually says about Noah's flood. These imaginative interpretations are the subject of the following chapter.

Chapter 8

Hyperbole, Allegory, Myth, or What?

Most people look to Genesis 6–9 for all the details about Noah's flood. However, other passages in Genesis, as well as portions of 1 Kings, Job, Psalms, Proverbs, Matthew, Luke, Acts, Romans, Colossians, Hebrews, and 2 Peter also speak of Noah and the flood (see chapter 12). These additional passages shed significant light on our understanding of the flood and its purpose in God's plan. However, because it's the earliest and most extensive narrative, theologians and others tend to debate most often and ardently over the Genesis 6–9 text, or what they think it says.

Interpretations of Genesis 6–9 that characterized the first 16 centuries of Christian scholarship were few and largely similar. Since then, however, these interpretations have evolved into a bewildering array of diverse scenarios. Now, in the twenty-first century, views on the text differ so widely that to find any points of agreement can be challenging. This glaring lack of consensus among today's Christian pastors and scholars about how to understand Noah's flood has caused many people to question whether they can trust anything that the Bible says. What follows is a brief survey and assessment of currently debated flood interpretations.

A Tranquil Flood

Biblically conservative Christian theologians and scientists now widely acknowledge the absence of credible scientific evidence for a recent, globally extensive flood that shaped or reshaped most of Earth's geological features. The lack of such evidence has led some scholars to adopt a "tranquil" or "small-scale" interpretation of the flood described in Genesis 6–9.[1] They conclude that Noah's flood was a local/regional catastrophe, one that impacted a relatively small percentage of the human population. For this targeted segment of humanity, Noah's flood would have appeared utterly catastrophic, of course, leaving no known (to them, at least) survivors. Scholars who interpret Genesis 6–8 as describing a brief, regional flood argue that it would leave behind no lasting geological evidence.

Christian scholars who believe Noah's flood impacted only a portion of the human population point out that, in the ancient world, several people groups became geographically isolated from other groups, sometimes for several generations. Examples would be the Māoris in New Zealand, the Inuit in northern Canada, and some indigenous peoples of the Amazon region and New Guinea. An isolated people group could, understandably, interpret a substantial regional flood as a catastrophe impacting the whole of humanity.

In a paper published in 2002 by *Perspectives on Science and Christian Faith*, geologist Carol Hill described Noah's flood as "a real, historical event that covered—not the whole world—but the whole of Noah's world,"[2] a world that did not, as she later explained, encompass all of humanity. In 2009, petrologist Lorence Collins suggested that the survivors of Noah's flood "lived in a village on natural levees on the lower parts of either the Euphrates or Tigris Rivers where the flood waters covered their village, natural levees, and adjacent flood plains for distances of 160 to 320 kilometers so that no land could be seen, and their 'whole world' would have been under water,"[3] but not the world of other people.

Noah's flood could be considered relatively "tranquil" in such scenarios, especially by contrast with a deluge thought to have drastically altered all Earth's geological features and destroyed a large, widely dispersed population. Genesis 6–7 is explicit, however, in declaring that *all* humans and *all* their animals except those aboard Noah's ark were wiped out by Noah's flood. The words "all," "every," and "everything" appear more than 40 times in Genesis 6–8. This repetition seems counter to the idea that Noah's flood impacted only a fraction, rather than the totality, of the human population (see chapters 10–14).

A Case of Hyperbole

Some Bible scholars have developed an interpretive model affirming that, while Noah's flood was a real historical event, the Genesis account is permeated with hyperbolic language. As the Biblical Archaeology Society stated in 2019, "a flood that seemed large and tragic to an ancient Mesopotamian village could easily grow over centuries into a Biblical cleansing of the entire Earth at the hands of God."[4]

No scholars doubt that the Bible authors made effective use of hyperbole as a literary device. An often-cited example is Mark 1:5, which says that "the whole Judean countryside and all the people of Jerusalem went out to him [John the Baptist]." John 1 verses 19 and 22 tell us some Jews in Jerusalem did not go out to hear John the Baptist's preaching in person. Clearly, Mark wants to emphasize the vastness of the crowd who heard John the Baptist's preaching—a crowd so large as to seem like the entire population of Jerusalem.

Theologians and ancient Near East historians Tremper Longman III and John Walton comment that Genesis 6–8 repeatedly uses hyperbole in describing the extent of Noah's flood, the people and animals destroyed by the flood, the floodwaters' sources, and the size of Noah's ark.[5] According to their view, the ark's dimensions as recorded in Genesis are impossibly large. To them it appears inconceivable that Noah could have built such a huge ship.[6] Thus, they surmise that nearly everything written in Genesis 6–8 must be taken as exaggeration. They conclude, "Though some modern readers don't see it, the original audience would have understood that such a description is hyperbole."[7]

One challenge to interpreting Genesis 6–9 as hyperbole would be the exactness of dimensions given in Genesis 6:15 for the ark's construction. The specificity seems inconsistent with hyperbole. Consider the difference between these two statements: "This luggage weighs a ton," and "This luggage weighs 74 pounds." The former is recognizable as hyperbole. The latter, given its numerical detail, would be understood as literal.

Not only does Genesis 6 offer precise dimensions for the ark (300 x 50 x 30 cubits), but it also gives specific instructions for the vessel's architecture (3 decks, a roof, a door, and a window along the length of the ark that measures one cubit tall). Such precise construction details also appear inconsistent with the use of hyperbole. Nevertheless, Longman and Walton approach the text with a firm conviction: Noah could not possibly have built such a large wooden ship with the technology at his disposal.[8] I believe this certainty warrants a challenge, even without invoking the miraculous (see chapter 14).

Other Christian scholars conclude that the Genesis 6–9 description must be hyperbolic because they assume the human population had already dispersed globally, or nearly so, by the time of Noah. No physical evidence supports a flood large enough to wipe out all humans and all their animals *if*, indeed, humans already occupied most of Earth's continents and islands. Furthermore, no reputable geophysicist or geologist would consider such a vast flood physically possible. Even if one were to invoke divine miracles to explain a flood of such enormity, the required intervention (including the provision and removal of sufficient water) without leaving a trace of physical evidence would appear to violate God's character (see chapter 13).

Symbolic/Allegorical Story

While nearly all Bible scholars of Christianity's first 16 centuries affirmed that the biblical account of Noah's flood held symbolic meaning (see chapter 4), they nevertheless regarded the account as historical. They expressed no doubt that its main message was a literal, historical one, a record of God's divine intervention in the world of his own making.

Common among many Bible scholars today is the view that Genesis 6–9 must be *strictly* symbolic or allegorical. Like Longman and Walton, they envision insurmountable conflicts with science that force them to "reject the historical-critical approach to the flood story."[9] In their book *The Bible Unearthed*, Israel Finkelstein and Neil Asher Silberman state that Genesis 6–9 portrays the mistaken worldviews of the ancient Hebrews and their Near Eastern contemporaries, not a true historical event.[10] In his book *Noah's Flood: The Genesis Story in Western Thought*, Norman Cohn provides an overview of modern western Bible scholars' and theologians' responses to Genesis 6–9 and notes consistent agreement that the text must be understood as purely symbolic, allegorical, or mythological.[11]

Nowhere in the Genesis 6–9 text do clues exist implying only a symbolic or allegorical interpretation is intended. Elsewhere in the Bible, authors explicitly clarify when their writing is *not* to be understood literally. For example, Jesus's parables documented in the four Gospels are clearly identified as parables. Jesus's audience always understood them to be parables. Jesus frequently explained the allegorical message embedded in the parable.

Bible authors certainly use symbols and symbolic references in their writing, but the use of symbols is sparing, not pervasive. When symbols are used, they are defined or explained as representations of something else. For example, in the book of Revelation, the Greek word for star, or stars, sometimes takes on a symbolic meaning, but the reader is told about the symbolic meaning. Revelation 1:20 says that stars can be symbols for angels.

In John 6:51, Jesus proclaimed to a crowd of Jews the symbolism of bread. He declared, "I am the living bread that came down from heaven." Later, at the Last Supper (Matthew 26:17–29; Mark 14:12–25; Luke 22:7–20), Jesus explained to his disciples that the Passover bread represents his body and the wine symbolizes his blood. It would be through his sacrificed body that his followers could stand faultless before God the Father and enjoy eternal fellowship with him. It would be through his shed blood that his followers would receive power from the Father to live according to his will.

In the book of Ezekiel (chapter 4), we read that God instructed the prophet to build a model of the city of Jerusalem with siege works, a ramp up to it, and battering rams around it in front of the Jewish leaders—completing all this work while unable to utter a word. The Jewish leaders all knew that what they had observed was an object lesson, but their pride kept them from grasping that its message was meant for them. Still, they remained curious and, at just the right moment, God restored Ezekiel's voice to deliver a stinging message.

Nothing like what is evident in any of these passages or similar passages appears in Genesis 6–9. The story is presented as historical narrative in the same sense as the stories in Genesis 12–50, Joshua, Judges, Ruth, 1 and 2 Samuel, 1 and 2 Kings, and 1 and 2 Chronicles. Theologians and other scholars and lay readers who reject a historical reading of Genesis 6–9 do so because such a reading *appears* to counter what's real or even possible. In chapter 12, I address familiar misreadings of Genesis 6–9. In chapter 13, I provide relevant scientific findings.

Mytho-history or Myth

Myth, as distinct from allegory or symbolism, comes from the Greek word *mythos*, meaning "story." For centuries the word referred to a traditional story without implying any judgment regarding truthfulness (or lack thereof). As used in today's culture, it typically refers to an untrue story or belief. When twenty-first century theologians identify Genesis 6–9 as myth, they typically reference the latter definition.

Christian philosopher and apologist William Lane Craig has responded to this development by using a hybrid term, "mytho-history." In his book, *In Quest of the Historical Adam*, Craig offers this mytho-history perspective on the early chapters of Genesis: "The primaeval narratives [Genesis 1–11] belong to the genre of myth principally on the basis of their sharing common mythic themes and their effort to anchor present realities in the deep past."[12] Craig considers these chapters as partly, though not entirely, mythical. Concerning Noah's flood, Craig writes, "In the flood story we are dealing not just with exaggeration but with the genre of myth."[13]

Craig explains his reason for rejecting Genesis 6–11 as an accurate account of history. He sees this portion of Scripture as filled with many "fantastic" elements: "nonmiraculous features of the story that, if taken literally, are palpably false."[14] Such elements include extraordinarily long lifespans, the Nephilim, the flood's extent (large enough to destroy all humans except those on board the ark), the ark's size, the localization of all humans and their animals in just a few contiguous regions of Earth's surface, the Tower of Babel, and the Table of Nations.[15] In his long section on fantastic elements, Craig critiques a mischaracterization of my model for Noah's flood (see chapter 15).

Theologian Peter Enns, among other Christian scholars, considers Genesis 6–9 as having little if any connection with actual history.[16] In the words of fellow theologian Michael Kruger, Enns approaches Scripture as if "filled with myth and rewritten stories,"[17] including the flood, simply "reworked to tell Israel's story."[18] In *The Bible Tells Me So: Why Defending Scripture Has Made Us Unable to Read It*, Enns expresses his belief that "storytelling is a better way of understanding what the Bible is doing with the past than 'history writing.'"[19]

Enns defines *myth* as "an ancient, premodern, prescientific way of addressing questions of ultimate origins and meaning in the form of stories: Who are we? Where do we come from?"[20] He considers Israel's flood story "a theological expression of self-definition."[21] Genesis 1–9, he says, is "not history in any normally accepted sense of the word today, . . . and most certainly not science."[22] He adds, "Genesis cries out to be read as something other than a historical description of events."[23] Rather, these chapters make "a theological statement in an ancient idiom."[24]

Theologian Kenton Sparks is just one among several who agree with Enns's assessment of Noah's flood as myth (by Enns's definition of the word).[25] Sparks brands both the Genesis creation and flood texts as "the recycling of Near Eastern myths and motifs."[26] He is far from alone in this judgment.

Alan Dundes (professor of anthropology and folklore at University of California, Berkeley and editor and coauthor of *The Flood Myth*[27]) agrees that Genesis 6–9 is appropriately labeled "myth" and that this passage is not to be taken as actual history. In the chapter he contributed to the volume, Dundes suggests that the Genesis 6–9 narrative is best understood as a cultural story intended to make moral and religious points.[28] The collection of essays also includes a chapter entitled "Noah and the Flood in Jewish Legend," written by Louis Ginzberg, a twentieth-century Jewish rabbi, historian, and Old Testament scholar.[29] Ginzberg describes Genesis 6–11 as a summary of Jewish legends and folklore about Noah and the flood.[30]

In their book *The Lost World of Scripture*, theologians John Walton and Brent Sandy argue that Genesis 1–11 should be appropriately identified (by genre) as "mythography," not historiography.[31] In their words, "though often adopting a narrative structure by recounting events, it [mythography] is generally not interested in those events as events that can be connected with the human world."[32] In other words, the biblical account of Noah and the flood has nothing to do with actual people and places and events.

Interpreting Genesis 1–11 as mythical rather than as historical narrative is a relatively recent development in biblical interpretation. Theologians promoting this shift essentially partition the book of Genesis into two parts: a primeval, mythical section (Genesis 1–11) and a patriarchal, historical section (Genesis 12–50). This partitioning raises an important question. Is this distinction warranted? Can it be justified from within the text of Genesis?

A case might be made for partitioning Genesis into two sections *if* one part were clearly patriarchal and the other, clearly not. However, both Genesis 1–11 and Genesis 12–50 fit the definition of patriarchal accounts. In fact, the reader finds more patriarchs and genealogies of patriarchs in Genesis 1–11 than in Genesis 12–50. These chapters of Scripture that provide a glimpse into the deep past are expressed in the same narrative style as Joshua, Judges, Ruth, 1 and 2 Samuel, 1 and 2 Kings, and 1 and 2 Chronicles. Ironically, most evangelical theologians who question or deny the historicity of Genesis 6–11 continue to affirm the truthfulness and historicity of these other narrative texts.[33]

Both Genesis 1–11 and Genesis 12–50 make frequent use of the Hebrew word *toledot*,[34] which in English translates as a phrase: "these are the generations of" or "this is the account of." *Toledot* appears six times in Genesis 1–11 (2:4, 5:1, 6:9, 10:1, 11:10, and 11:27) and five times in Genesis 12–50 (25:12, 25:19, 36:1, 36:9, and 37:2). This frequent use of *toledot* in both parts of the book strongly suggests that all of Genesis is to be interpreted as a single patriarchal narrative. With respect to Noah's flood, *toledot* appears in the introduction of the flood account, Genesis 6:9, explicitly connecting chapters 6 through 9 with the rest of the book of Genesis.

Theologian and Old Testament scholar Walter Kaiser observed that Genesis 1–11 contains "64 geographical terms, 88 personal names, 48 generic names and at least 21 identifiable cultural items."[35] Such frequent and diverse references to names, places, and cultural items, Kaiser concluded, indicates that Genesis 1–11 is intended to be interpreted as a literal, historical narrative, not as a myth.[36] To separate the profound theology of Genesis 6–11 from its history requires the imposition of external material, specifically naturalistic scientific models. Genesis 6–11 read on its own terms, without this external bias, clearly is *not* something other than historical theology.

Apparently, the motivation for interpreting Genesis 6–9 as mythical rather than as historical is the assumption that a literal, historical interpretation can't be defended in light of established facts. Tremper Longman III and John Walton concede as much when commenting on the elements of Noah's flood. They assert, "The Bible is not *providing* scientific revelation; it is being *silent* on scientific matters" (emphasis in the original).[37] (See chapter 12 for discussion of misunderstandings about a "literal" reading of Genesis 6–9 and chapter 13 for treatment of relevant scientific concerns.)

Examples of Accommodation

Twenty-first century theologians who hold to a nonliteral, nonhistorical interpretation of Genesis 1–11 share a common interpretive stance. Each engages in some form of "accommodation" while, at the same time, declaring belief in the divine inspiration of Scripture.[38] They say the Holy Spirit guaranteed and preserved the complete inerrancy of the Bible's *primary* theological message while tolerating, or "accommodating," the human authors' mistaken ideas on matters of theological insignificance. Some of the matters they deem theologically insignificant include those with relevance to science, history, and geography.

Accommodationists acknowledge no need to interpret Genesis 6–9 as an accurate historical account. They encourage readers of Genesis 1–11 to ignore all statements touching upon science and history and, instead, glean from the text its primary theological message. Enns and Sparks claim their approach to Bible interpretation is the *only* way to defend the reliability and authority of biblical revelation. As Sparks says, "Accommodation does not introduce errors into Scripture; it is instead a theological explanation for the presence of human errors in Scripture."[39]

The accommodationist approach advocated by Enns and Sparks, among others, collides head-on with explicit biblical statements, and I hope to show that it's not necessitated by the facts of history and/or science. As Jesus declared, "Scripture cannot be broken" (John 10:35, ESV). Matthew 5:18 records his comment that "not the smallest letter, not the least stroke of a pen, will by any means disappear from the Law [Old Testament] until everything is accomplished." The apostle Paul, in his second letter to Timothy, writes, "All Scripture is *God-breathed* and is useful for teaching, rebuking, correcting, and training in righteousness" (emphasis mine, 2 Timothy 3:16). In his first epistle, the apostle Peter affirms, "No prophecy of Scripture came about by the prophet's own interpretation of things. For prophecy never had its origin in the human will, but prophets, though human, spoke from God as they were carried along by the Holy Spirit" (2 Peter 1:20–21).

If numerous scientific, historical, and geographical errors were proven to exist in the biblical passages recounting Noah's flood, then the historic doctrine of biblical inerrancy (belief that the Bible is inspired and without error in everything it addresses) would no longer hold. A seismic shift of immense significance would be in order. A change of such magnitude warrants a thoughtful pause and scrupulous investigation.

Credibility challenges to Genesis 6–9 come not only from doubts about what's physically possible but also from the noticeable similarity between the biblical flood story and the Mesopotamian epics. Extant manuscripts and other references to Mesopotamian legends predate the biblical writings, and this combination of timing and similarity has caused many scholars to consider Genesis 6–9 as a "reworked" version of the Mesopotamian stories of a flood that destroyed all people except those who escaped in a large boat. The next chapter addresses these Mesopotamian flood stories and offers a detailed comparison and contrast between Genesis 6–9 and the Mesopotamian story that comes closest to matching it: the Epic of Gilgamesh.

Chapter 9

The Genesis Flood and the Mesopotamian Epics

Historians refer to Mesopotamia (see figure 9.1) as the cradle of civilization with good reason. Mesopotamia is where the first known urban settlements (cities) formed, where nomadic peoples are first known to have engaged in intensive and extensive agriculture, and where earliest known architectural monuments and temples were constructed. Mesopotamia saw society's diversification into specialized industries and trading systems. Such developments gave birth to written language, works of literature, and historical and accounting records.

In the ruins of Mesopotamia's ancient cities, archaeologists have discovered three distinct written epics (narrative poems recounting heroic deeds) of a flood that wiped out all people other than a few individuals on board a large boat. Among the hundreds of ancient flood stories and myths, Eridu Genesis, the Epic of Atra Hasis, and the Epic of Gilgamesh come relatively close to matching the account in Genesis 6–9. As *written* accounts, they predate Genesis 6–9. Based on this timing, many historians reason that the flood story recorded in Genesis may not be the original. They speculate that what is recorded in Genesis 6–9 is either a corrupted version or a reworked version of a more ancient catastrophic flood story. Such speculation calls for careful examination of these three Mesopotamian epics.

Eridu Genesis

The oldest of the flood narratives was written by the ancient Sumerians. Popularly known as Eridu Genesis, it combines accounts of creation and a vast flood. It was found on a single tablet excavated at Nippur and dated to circa 1600 BC.[1] The beginning of the tablet is missing, and other parts are damaged,

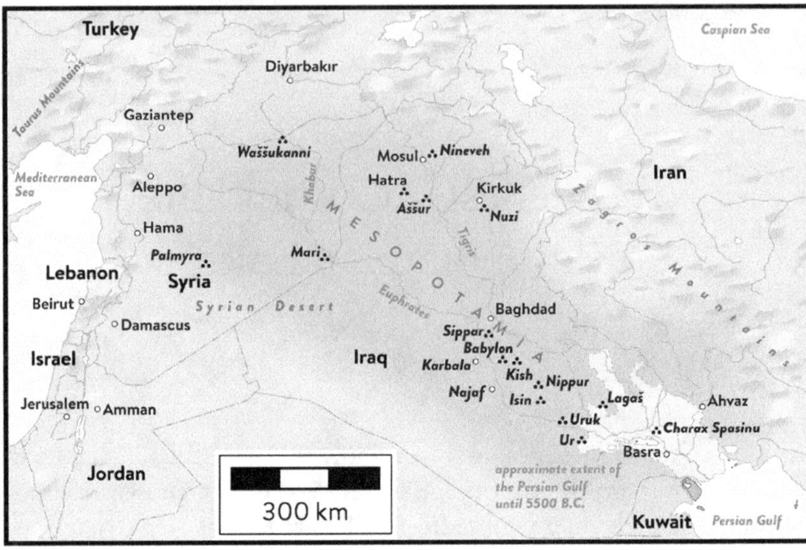

Figure 9.1: Mesopotamia, Showing the Sites of Ancient Cities and the Ancient Shoreline of the Persian Gulf
Credit: Goran Tek-en, Creative Commons Attribution

but much is still readable.

According to the Eridu Genesis, a team of four gods fashioned humans from clay for the purpose of cultivating the ground, building cities, caring for flocks, and establishing worship of themselves (the gods). For reasons missing from the undamaged parts of the tablet, the gods became displeased with the humans and decided to send a massive flood to wipe them out. As the story goes, one of the gods, Enki (the god of seas, lakes, and rivers), warned Ziusudra, a humble and obedient priest and king, of the impending flood. Although what follows includes no mention of a boat or its construction, most ancient Near Eastern scholars presume that the damaged, unreadable portion of the tablet once included Enki's instructions to Ziusudra to build a boat.

The next part of the tablet describes the flood. It says that a terrible storm raged for seven days and nights, tossing about the boat on which Ziusudra and a few companions rode out the storm and the ensuing flood. When the sun came out, Ziusudra sacrificed sheep and oxen. An ancient Sumerian tablet, known as the Epic of Ziusudra, says that with the subsiding of the flood, Ziusudra's boat floated down the Euphrates River into the "great waters" (most likely a reference to the Persian Gulf).

Epic of Atra-Hasīs

Of all the ancient nonbiblical stories of a deadly flood, the Epic of Atra-Hasīs offers the most extensive description of the entire event.[2] The account was written in the Akkadian language on three sets of tablets, two by Assyrian scribes and one by a Babylonian scribe. The three sets of tablets date back to the seventeenth century, BC.

Like the Eridu Genesis, the Epic of Atra-Hasīs deals with both the creation of human beings and a subsequent devastating flood. The Epic of Atra-Hasīs depicts two sets of gods, the Anunna (great gods) and the Igigi (lower gods). The Anunna subject the Igigi to backbreaking work. The Igigi start to complain and eventually revolt against their forced labor. In negotiations between the Igigi and the Anunna, the Anunna agree to create humans in order to relieve the Igigi of their arduous labor. The gods slaughter one of their own and mix the flesh and blood of the slaughtered god with clay to make the humans.

The human population grows and, with it, the noise of people's chatter. The racket causes the gods to lose sleep and so irritates them that the Anunna decide to eradicate the humans with a great flood. However, one of the gods warns a human king, Atra-Hasīs, in a dream. This same god explains the dream to Atra-Hasīs, speaking to him from the walls of his house. He tells Atra-Hasīs to flee his house and build a boat. The boat is to have a roof over its entire length so that no sunlight reaches inside the boat. The boat is to be constructed of reeds and wood and coated with firm pitch.

As the story goes, Atra-Hasīs takes his family, sheep, cattle, and birds on board the boat. The flood comes with such force and darkness that people no longer recognize one another. Everyone outside Atra-Hasīs's boat is killed by the water, wind, clamor, and darkness, but Atra-Hasīs and those with him survive. After the flood, the gods limit the human birth rate and lifespan as means of keeping the noise of human chatter down to a tolerable level.

The Epic of Gilgamesh

The most widely known of the Mesopotamian flood stories, apart from the biblical account, is the Epic of Gilgamesh. It is recorded on 12 Akkadian tablets found by archaeologists in Nineveh in the library of Assyrian king

Ashurbanipal. This incomplete set of tablets dates to about 650 BC. Portions of some of these tablets date to about 1500–1200 BC. Near Eastern scholars tend to agree that the tablets and portions are copies of earlier originals that may date back to about 2100–1800 BC. Tablet XI contains the story of a universal (with respect to humanity, not geography) flood.[3]

According to Tablet XI, a council of gods determined to eliminate humanity by sending a huge, catastrophic flood upon the earth. One of the gods, Ea, chose to warn Utnapishtim, an exceptionally wise human leader, of the impending disaster. Ea then instructed Utnapishtim to tear down his house and build a large boat.

According to the story, the finished boat, made of reeds and wood, measured 120 cubits long, wide, and high (whether cubical or cylindrical in shape, the tablet does not specify.) Pitch, bitumen, and plugs were employed to prevent leaks. To manage construction of the boat's six decks, seven levels, nine compartments, and roof, Utnapishtim recruited a large team of reed workers, carpenters, porters, and children. They were able to complete the vessel's construction in just six days, says the tablet.

Utnapishtim then loaded the boat "with all his kith and kin" and "all the beasts and animals of the field and the craftsmen,"[4] as well as provisions for the animals and people and all his silver and gold.[5] At this point the gods sent the flood, so "frightful to behold"[6] that it terrified even the gods,[7] who are said to have been "cowering like dogs."[8]

During the flood, the days turned as black as night.[9] Outside the boat, "Land shattered like a . . . pot. All day long the South Wind blew . . . blowing fast, submerging the mountain in water, overwhelming the people like an attack. No one could see his fellow, they could not recognize each other in the torrent"[10] (ellipses indicate text missing from the tablet). The storm is said to have flattened the land.[11]

After six days and seven nights the floodwaters abated. Utnapishtim noted that the terrain was "as flat as a roof,"[12] except for Mount Nimush, where the boat had firmly lodged.[13] On the seventh day, Utnapishtim released a dove that came back to him. He then released a swallow that also came back. Finally, he released a raven that didn't return. Thereupon, Utnapishtim had all the people and animals disembark from the boat.

After the disembarkation, Utnapishtim reportedly "sacrificed" and "offered incense in front of the mountain-ziggurat,"[14] burning reeds, cedar, and myrtle for the incense. Not all the gods, however, were placated by Utnapishtim's sacrifice and incense. Enlil became furious that some humans survived the flood.

The tablet ends with Utnapishtim informing King Gilgamesh about a special plant, "a plant against decay."[15] "The plant's name [was], 'The Old Man Becomes a Young Man.'"[16] Utnapishtim never got to eat of this plant of life because "A snake [that] smelled the fragrance of the plant, silently came up and carried off the plant."[17]

A Rewrite of the Legend?

Many skeptics of Christianity and even some Christian scholars of the Old Testament[18] argue that the Genesis account of Noah's flood may be a reworked version of these ancient Mesopotamian flood legends. The perceived similarities are striking, indeed, and they're worth reviewing, but there's much to consider besides commonalities.

First, here is a list of story features the Eridu Genesis, Epic of Atra-Hasīs, and the Epic of Gilgamesh do share with Genesis 6–9:[19]

1. A fearsome flood kills all humans but those few chosen by the human hero.
2. A god or gods warn the hero of an impending destructive flood.
3. The hero is a uniquely admirable character.
4. A god or gods tell the hero to build a large boat.
5. The boat is sealed with pitch.
6. The hero and a few people of his choosing survive the flood in the boat.
7. Once the flood subsides, the hero and the people with him exit the boat and restart civilization.

The Epic of Atra-Hasīs and the Epic of Gilgamesh share two additional features with the Genesis text: (1) the boat has a roof that runs the vessel's full length; and (2) after leaving the boat, the hero offers an animal sacrifice. The Epic of Gilgamesh shares several additional features in common with Genesis 6–9, including these:

1. A god or gods give specific instructions to the hero concerning the dimensions and/or architecture of the boat.
2. The boat has multiple levels or decks.
3. The boats are roughly equal in volume.
4. A god or gods tell the hero to take certain animals on board.
5. The hero takes adequate provisions for fellow passengers and animals.
6. When the floodwaters abate, the boat comes to rest on a hill or

mountain.

7. The hero releases birds to determine whether the floodwaters have receded.

While these overlapping details reflect a connection between the epics and the biblical account, there's much more to consider in identifying the stories' source. Scholars who argue that Genesis 6–9 is a reworked version of older Mesopotamian flood legends, especially the Epic of Gilgamesh, fail to adequately consider major factors that weigh against their conclusion.

Common and Uncommon Features

The first factor is that the Mesopotamian flood legends have much more in common with one another than they do with Genesis 6–11. For example, the Eridu Genesis, Epic of Atra-Hasīs, and Epic of Gilgamesh all reflect primitive polytheism. What's especially telling is the stark contrast between the human-like gods depicted in the epics and the morally perfect God of the Bible. The Mesopotamian gods engage each other in serious conflict and deception, unlike the triune God of the Bible—who is singular in essence and purpose.

Obviously, the differences between Genesis 6–9 and the Mesopotamian flood legends, even in the case of the Epic of Gilgamesh, are far greater and more significant than the similarities. The dimensions, architecture, and building material of Noah's ark are notably different from those of Utnapishtim's boat. Noah's ark is constructed of hardwood. Utnapishtim's boat is made of reeds and wood. While Noah's material has tensile strength sufficient for a seaworthy vessel measuring as large as 300 x 50 x 30 cubits (see chapter 14), a combination of reeds and wood would not, and certainly could not sustain a vessel measuring 120 x 120 x 120 cubits. Further, the dimensions and shape of Utnapishtim's boat would render such a vessel unseaworthy. The dimensions and shape of Noah's boat match those of modern ocean-crossing ships. Noah's ark was wholly sealed with pitch. Utnapishtim's boat is sealed with pitch, bitumen, and plugs—far less reliable sealants.

Only eight humans enter Noah's ark. All are members of Noah's immediate family. Utnapishtim's boat includes all Utnapishtim's kith (unrelated friends and associates) and kin, plus all the craftsmen that helped construct his boat. Genesis 6–8 specifies the kinds and numbers of animals on board Noah's ark (see chapters 10, 12–14). The Epic of Gilgamesh simply states that Utnapishtim took on board "all the beasts and animals of the field." Genesis 6–9 makes no mention of Noah or his family members taking on board any precious metals.

In the Epic of Gilgamesh, Utnapishtim takes on board all his silver and gold.

In the Epic of Gilgamesh, the flood lasts only six days and seven nights and yet somehow reshapes the terrain all around it, flattening all but one lone mountain. Genesis 6–9 makes no mention of Noah's flood having changed any of Earth's geological or geographical features, even though the rains fell much longer than six days, and the floodwaters remained at full height for 150 days, requiring more than a year to fully recede.

Many more differences could be cited, from the types of birds released to the types of sacrifices made once the people and animals were able to leave their boats behind. The Gilgamesh's mention of a ziggurat represents what can rightly be described as a major embarrassment. The raging flood described in the Epic of Gilgamesh would have destroyed any previously existing ziggurat or ziggurat-type structure in the region. To have rebuilt such a structure immediately after the flood also would have been unimaginable. Ziggurats are massive brick structures 50–150 meters (165–500 feet) high consisting of huge multiple stories with a temple on top. They obviously require a long time to build, even when unlimited human labor and material resources are available. Yet, the Epic of Gilgamesh implies that Utnapishtim burned incense in front of a ziggurat soon after disembarking from his boat.

As mentioned earlier, among the council of gods in the Epic of Gilgamesh, Enlil was furious that any humans had survived the flood. He was displeased with Utnapishtim and his sacrifice. In Genesis 8–9, however, we read that God was pleased with Noah and accepted his sacrifice. He then blessed Noah and his family.

Perhaps the greatest difference of all between Genesis and the Epic of Gilgamesh would be identified as the contrasting motivation for sending the flood. The gods in the Epic of Gilgamesh cause a devastating deluge not to protect humanity, but to destroy humanity for no clear reason. Their action is arbitrary. Once the flood is underway, the gods are said to flee in terror to some distant upper reaches, where they crouch in fear. Then, suddenly, the flood ends. The biblical story reveals a God who acts out of grief and necessity to rescue humans, his beloved creatures, from their sin and from destroying each other and the rest of creation. God's actions reflected both the pain and purpose of a loving heart.

Cultural Agenda

Another distinguishing factor is that the Mesopotamian flood legends affirm a particular cultural agenda that's missing from the Genesis story. The

Mesopotamian flood epics reflect and affirm a rigid hierarchy within the human population. A few kings and priests rule over the rest of the human population,[20] the vast majority of whom are serfs. These underlings exist to do heavy labor and pay onerous taxes so that the few kings and priests can live comfortably as the cultural elite. The serfs are second-class citizens who exist to serve the ruling class.

There is no such portrayal in Genesis. God declares in Genesis 1 that all humans bear his own image, equally. Each person has equal standing before God and before every other person. In the Genesis account of the flood, as in the rest of the Bible, all humans are of equal worth before God, and rescue is available to all. Although there was a period in the history of the Hebrew people when they were ruled by kings, this social structure was not God's plan. God granted them a king only in response to their pleading for one (see 1 Samuel 8:1–22). Nevertheless, Israelite kings, unlike the kings of surrounding nations, were subject to laws and constraints. Frequently, God sent prophets to rebuke Israel's kings, both those who worshiped him, such as David (see 2 Samuel 12:1–23), and those who defied him.

Oldest Flood Account

The weightiest factor arguing against the view that Genesis 6–9 is, or could be, a reworked version of the Mesopotamian flood legends resides in assumptions about the time of their origin. The oldest written version of the flood story, as with any ancient story, may not necessarily be the original. All human languages first develop as oral expression. The written language takes form over time, if ever.

A substantial body of biblical evidence, supported by scientific findings, indicates that Noah's flood predates—by as much as many thousands of years—development of humanity's first written languages (see chapter 11). The Genesis account of the flood, as well as of the creation story, would have been transmitted and preserved orally for a very long time.

People who lack a written language have been known to preserve the important parts of their history through word-for-word memorization. Such societies prioritize training in precise memorization. Individuals who show special adeptness in oral memorization are singled out and chosen for special service as preservers of their society's most important records.[21] With the development of written languages and, in the twenty-first century, of data storage devices and cell phones, removing the need for lengthy, accurate memorization, this skill has become rare.

For a biblical example of oral preservation, we can look to the book of Job. The geographical places mentioned in the book suggest that Job lived before the nations of the Middle East had developed and expanded beyond city-states. The portrayal of Job, a patriarch, not a Levitical priest, performing animal sacrifices implies that the debate between Job and his friends occurred before the time of Moses and establishment of the Aaronic priesthood. The consensus among conservative Bible scholars is that Job lived in early patriarchal times, likely before the time of Abraham.[22]

The literary structure of the book of Job is unique among the books of the Bible. In fact, it serves as the classic example of the "epic poem." This form is recognized as a powerful aid to memorization, thus to oral transmission. Not until the time of Moses did Hebrew become a written language. Thus, for the book of Job to have been included in the canon of Scripture, it was deemed to have been faithfully preserved through oral transmission. The silence of the young man, Elihu, until the thirty-second chapter of Job leads to viable speculation that he may have chosen (or have been chosen) to record the momentous conversation among Job and his friends.

To acknowledge that Genesis 6–9 is not the oldest *written* record of an ancient flood event says nothing conclusive about the timing of the event. Hebrew became a written language no earlier than the mid-fifteenth century BC, the most likely date for Moses's leadership of the Hebrews' exodus from Egypt and the writing of the Torah (Genesis, Exodus, Leviticus, Numbers, and Deuteronomy). As already noted, the earliest extant record of the Eridu Genesis, the Epic of Atra-Hasīs, and the Epic of Gilgamesh are dated to circa 650 BC, though nearly all ancient Near Eastern scholars consider these tablets copies of earlier originals, dating back to about 1500–1600 BC. While these dates are earlier than 1450 BC—the most likely date for the writing of the Torah—they are earlier by only a small margin and offer no clear evidence of the timing of the events they depict.

A Reasonable Conclusion

Notable variations among the three Mesopotamian flood legends suggest that they likely derived from a few different cultural (and oral) sources. Moses, on the other hand, relied not solely on orally transmitted stories but also, according to the Torah, on direct communication with God. By the power and presence of the Holy Spirit, God oversaw and ensured the accuracy of what Moses recorded in the Torah (see Exodus 24:4; Numbers 12:6–8; Deuteronomy 18:15; Acts 3:22).

Secondary support for the divine inspiration of Moses's account of Noah's flood and its aftermath comes from its frequently and unjustifiably maligned scientific implausibility. In this regard, Genesis 6–9 holds up where the Mesopotamian legends do not. Consider the stability and seaworthiness of Noah's ark, a stark contrast to Utnapishtim's boat (see chapter 14). Consider the Gilgamesh rains that continued for only one week and abated just as quickly thereafter. If a flood formed at all from such a limited downpour, however ferocious the downpour, it would not have been sufficient to destroy either most or all of humanity outside the boat.

The quantity of cargo and passengers represents another challenge. Utnapishtim's boat, with a volume roughly the same as Noah's ark, had many more passengers and much more cargo. It included all Utnapishtim's kin, associates, and fellow craftsmen, plus animals (of unstated distinction or limitation), provisions, the seed of all living things, and some quantity of gold and silver.

The greatest difference of all between the Genesis account and the Mesopotamian stories lies in what they reveal about the One or ones, and the "what," responsible for the flood. The gods of the Mesopotamian legends are clearly human constructs. They manifest distinctly human character attributes and no redemptive purpose in their actions and decisions.

The Bible stands apart from these epics in its core message, as well as significant details. Though it may have been written at a later date, the oral can reasonably account for its preservation. Thus, the claim that the biblical narrative borrowed from these epics appears questionable, at best. The Genesis narrative of Noah's flood proves historically and theologically credible whereas the Mesopotamian flood epics do not.

The chapters to follow address the main challenges to the scientific credibility of the biblical account of Noah's flood—its extent, timing, duration, and more. The next chapter explores the deeper question, a profoundly theological one, about the divine purpose for and scope of the flood's devastation.

Chapter 10

God's Response to Reprobation

The "facts" of the Genesis flood often are misunderstood by skeptics and, as a result, dismissed as fable. The bigger challenge, one that elicits an even stronger reaction from a wider audience, is the flood's theology—what people tend to think Genesis says about God's character. Even those who would defend the book of Genesis and all other books of Scripture as divinely inspired often struggle, mightily, to understand how the level of devastation reported in the flood story can be reconciled with the character of an all-loving God. How could a good and merciful God destroy all but a handful of people—clearly imperfect ones, at that—and so many innocent animals, creatures of his own making?

What seems a glaring inconsistency with respect to God's goodness and love deserves attention and must be addressed before attempting to unravel any other troubling aspects of Genesis 6–11. How is the biblical God's act of divine judgment any different from the Mesopotamian gods' anger? To deal with this all-important question requires an in-depth examination of God's attributes alongside a realistic consideration of context: first, the level of moral degradation to which those destroyed in Noah's flood had fallen and, second, the impact of this degradation on their animals and environment.

God's Attributes

As expressed in chapter 2 of *The Westminster Confession of Faith*[1] and affirmed in article 1 of the Belgic Confession,[2] God as revealed in the Bible and self-disclosed in nature is more than an all-loving, all-benevolent Being. The biblical God, the only God, is the transcendent source of all being and knowing, the embodiment of life, love, and truth. He is holy, "set apart" from all other beings, perfect in righteousness, wisdom, justice, mercy, and power—infinitely so. Not one of his attributes exists apart from any other.

God's holiness means God is wholly pure. This moral perfection implies God can't tolerate the presence of sin and evil. For God's love to be fully expressed and received, sin and evil must be dealt with and removed. God does not want our sin to be a barrier to his love.

God cannot overlook metastatic evil without violating his own character. God's holiness means he must take protective action. Nevertheless, when we see references to God's "wrath" in the Bible, we instinctively recoil. Why? We imagine his "wrath" resembles humanlike anger, an emotion that flares in response to wounded pride, careless or deliberate injury, defiance, loss (or lack) of control, or thwarted desire. However, as theologian Timothy Keller reminds us, "God only gets angry at the evil destroying the things he loves—his creation and the human race he made for his own glory and for our happiness."[3] To think God must be *either* a God of love *or* a God of wrath is a serious misunderstanding. Again, Keller reminds us, "*He is both, and if your concept of God can't include both, it will distort your view of reality in general and of forgiveness in particular*" (emphasis in the original).[4] God will not allow sin to overturn his redemptive plan.

God's gifts of forgiveness and mercy come at a price because God is just and holy, and evil is consequential. The Creator of the universe chose to pay the price, himself, for humanity's sin. He came to Earth in human flesh, lived without sinning, and offered his own perfect life as the one sacrifice that could possibly atone for all of humanity's sin—a sacrifice available to everyone by faith—a free-to-us, costly-to-him gift from God. This is the good news the Bible proclaims. Old Testament believers put their trust in God's promise to Abraham (Genesis 15), just as New Testament believers put their trust in God's fulfillment of that promise in Christ (Galatians 3:29).

Human Sin and Its Damage

When Jesus of Nazareth came to live among us as God in human form, he devoted much of his teaching to clarifying what our *sin* is. Most people who heard his teaching missed the message. Most people today still miss it. Jesus explained that in breaking any one of the Ten Commandments, we have broken all.[5] If we break a commandment in our mind but refrain from breaking it in deed or word, we're still guilty of breaking his commandments.[6] We have demonstrated our true nature, the same nature expressed by the first humans, Adam and Eve. We may think we would have chosen differently, chosen not to eat of the forbidden tree, but our own thoughts betray our imperfection even when we view ourselves as capable of perfect righteousness. As Scripture says, "all have sinned" (Romans 3:23). "There is no one righteous, not even one" (Romans 3:10), apart from Jesus Christ.

Sin describes the condition of our spirit,[7] the "self" we possess from the time we're conceived.[8] It's our tendency to live as if we're autonomous beings, rather than as creatures subject to our holy and loving Creator. *Sins*, plural, are the acts, words, and thoughts that reflect our *sin*, singular, the condition that characterizes our innermost being.

Sin's Damage

Our sinful state leads to sins that always lead to harm, even if (seemingly) only to the one who commits them.[9] Sin "defiles" the one who sins and impacts others, though not all sins damage us and others to the same degree or to the same extent.

The apostle Paul, in his first letter to the church at Corinth, writes that some sins do more injury and are more "defiling," than others.[10] Paul explains that sins committed against the body, such as murder, assault, fornication, adultery, and rape, do deeper and more enduring harm than sins outside the body, such as stealing, lying, cheating, slander, and deviousness.[11] All sins bring consequences, harming both the perpetrator and the one(s) victimized. All sins offend and grieve God.

Some sins are so damaging as to impact not only the victims and person(s) committing them but also the people and soulish animals in close contact with them (see sidebar *Nephesh* (Soulish) Animals). On some occasions, even the material possessions of the sinning person(s) can pose a danger to others.

The impact and spread of sin's defilement resembles the impact and spread of a deadly contagion. The Bible describes sin's effects as expanding outward in this order:

1. to the sinning person (Romans 7:8–11);
2. then to their spouse, progeny, and close associates (Exodus 20:5);
3. then to their soulish animals (Joshua 6:21);
4. then to their material possessions (Numbers 16:23–33); and
5. lastly to the land they inhabit (Leviticus 18:24–28).

Not since the days of Pentecost, when God poured out the Holy Spirit upon his followers and came to permanently dwell within them—Jews, Samaritans,

Nephesh (Soulish) Animals

The Hebrew verb to create something wholly new, *bara*, appears only three times in the Genesis 1 creation account. In verse one it refers to God's bringing into existence the physical creation, "the heavens and the earth," the entire physical universe. The next time we see this verb, it is used with reference to animals that are both physical and "soulish," the *nephesh*.[12]

Nephesh (mentioned 754 times in the Bible) are animal species that care for and nurture their offspring and/or mates. All bird and mammal species are included among *nephesh*. A few reptile species may also qualify. *Nephesh* are more than physical bodies with nervous, digestive, respiratory, circulatory, and reproductive systems, as marvelous and intricate as those systems are. *Nephesh* also manifest a degree of intelligence, intention, and emotion—what the ancient Hebrews would call soulish attributes.

Nephesh seem drawn to relate to us humans, especially to please us and/or assist us in some way. They have the capacity to form relational bonds with us, receiving and giving affection. They can be tamed by us, even trained to do our bidding. Because of their strong desire to please the person(s) to whom they are bonded, they tend to reflect the demeanor they're most familiar with. Consequently, they can be damaged, even irreparably and dangerously, by human sinfulness.

and Gentiles (Acts 2, 8, 10)—has the world seen the depths and extent of defilement, of wholesale depravity, as occurred in the days of Noah. The "salt of the earth" (Matthew 5:13), which Jesus said his disciples would become, is a spiritual preservative protecting cities and nations from whole-population reprobation (pervasive, continual, extreme, unrepentant evil) that occurred on rare occasions during ancient times. Two episodes are reported in Genesis, the first at the time of Noah and the second during Abraham's lifetime, specifically in the vicinity of Sodom and Gomorrah.

This second example of pervasive, rampant wickedness, or "reprobation," as the English Bible refers to it, provides a sense of the horrors associated with such degradation. Genesis 19:4–10 reports that the entire male population of Sodom, young and old, came to Lot's house intent on committing rape against the two male visitors Lot had taken in as guests. Lot's unholy ploy to prevent this attack temporarily held the crowd at bay but, at the same time, revealed the loss of his own integrity as a father.

To prevent the malignancy of Sodom's irreparable sin from spreading to nearby regions, God found it necessary to act, as would a cancer surgeon, by removing every trace of the deadly disease. Even the animals and possessions of the Sodomites had to be destroyed. To this day, the land that once was Sodom remains barren.

Not since the outpouring of the Holy Spirit at Pentecost has the world seen an entire nation descend to the depths of Sodom's wickedness. Certainly, there have been evil leaders, some alive today, who have demonstrated a monstrous degree of evil, but not an entire population. Each has been restrained, at least to some degree, against their will. The "salt of the earth" is still present.

Some people find it hard to believe that the God of the Old Testament is the same God revealed in the New Testament. Old Testament accounts of large-scale destruction ordered by God seem incompatible with God's expressions of mercy and love in the New Testament. Perhaps people fail to recognize God's instances of judgment as comparable to the work of a masterful surgeon who works skillfully and compassionately to remove a malignant tumor that threatens to take someone's life. God, like that surgeon, preserves and protects all that can be safely retained, taking out no more and no less than necessary.

God has not changed, but circumstances have, to a significant degree. The arrival and continuing presence of the Holy Spirit permanently indwelling the followers of Jesus Christ makes a clear difference.[13] Widespread respect for good over evil, even where confusion exists about which is which, still exerts a powerful influence throughout the world.

Dangers of Reprobation

The reprobation that plagued preflood society developed and spread over time. Romans 1:18–32 provides an overview of how reprobation develops and gains momentum. Verse 32 describes those who so persistently refuse to turn from evil that they no longer care to turn:

> Although they know God's righteous decree that those who do such things deserve death, they not only continue to do these very things but also approve of those who practice them.

Reprobate people are addicted to evil. They continually think, speak, and act perversely. Whatever the Bible declares good and true they consider evil and false, and vice versa. They cheer for those who do evil and speak lies. They enjoy recruiting others to think and act as they do. According to the apostle Peter,

> They never stop sinning; they seduce the unstable. . . . By appealing to the lustful desires of the flesh, they entice people who are just escaping from those who live in error. They promise them freedom, while they themselves are slaves of depravity.[14]

Anyone steeped in depravity for sufficient time will eventually be saturated. Moral corruption will at some point permeate their soul. The apostle Paul described such people as Satan's captives.[15] (Satan, as the Bible tells us, is the most powerful angel God created and the first of God's creatures to reject God's authority, love, and goodness.) Reprobate people appear to share Satan's desire to drag as many others as possible into captivity with them.[16]

God's wrath takes action to release those who've been taken captive by Satan. God moves to prevent them from continuing, irretrievably, in that direction by removing those already in irretrievable captivity. God knows when and how to wield his holy scalpel. He knows precisely how much corruption must be removed. He's an enemy to spiritual villainy and a friend to spiritual vitality. He operates from the motive of love to save as many people as possible from total moral collapse and the spiritual death that follows.

Preflood Depravity

In the years prior to Noah's time, we encounter a descendant of Cain named Lamech. The Bible records his bragging words: "I have killed a man for wounding me, a young man for injuring me" (Genesis 4:23). He goes on to threaten, "If Cain is avenged seven times, then Lamech seventy-seven times" (Genesis 4:24). By the time "human beings began to increase in number on the earth" (Genesis 6:1), depravity had increased all the more, to such a degree that it threatened to contaminate everyone, even people's soulish animals.

The Hebrew text describes such evil using two words: *shachat* and *hamas*. Lexicons define the former as "morally putrid, totally decayed, spiritually gangrenous, destroyed and wasted";[17] and the latter as "seeking to gain through assault, physical attack, cheating, and/or oppression."[18] No other society referred to in Scripture warranted the use of such strong language. Such pervasive evil was, and remains, unparalled throughout human history (although the New Testament warns that society will have returned to a similar condition at "the time of the end"[19]). It seems clear that the pride and violence expressed by Lamech had spread like a cancer and taken a horrible toll.[20]

The (potentially) long lifespans of people living before the flood (see the Genesis 5 genealogy and Genesis 6:3) had apparently favored the spread of violence. Serial killers had much more time and, therefore, many more opportunities to commit murder. No doubt the percentage of murderers in the population increased as their victims, likely the more righteous among them, died. It seems mathematically plausible that in the days before Noah's flood, by far the most common cause of death must have been murder.[21] Humanity was in imminent danger of self-extermination. In this context, Noah's flood was an expression of God's mercy, a rescue mission.

We are not told why God allowed conditions to reach such a deplorable state, but one possible reason comes to mind. God may have wanted to demonstrate to future generations that we humans lack the spiritual resources to save ourselves from sin, to overcome evil with good. God certainly did demonstrate, indelibly, that we need him to do for us what we can't do for ourselves. Perhaps he also wanted to persuade us that short lifespans benefit us under present conditions, as we await his ultimate return and deliverance from the presence and possibility of sin. We don't need more than a maximum of about 120 years (Genesis 6:3) to discover who God is and how to receive the abundant, eternal life he offers to repentant sinners.

Genesis 6 tells of God's agonizing grief over humanity's corrupt condition. With a heavy heart he cleansed the world to keep it from utter ruin. He found only one man who, with his family, could keep humanity from total annihilation. The story is all about rescue and mercy. God saved 100 percent of the nonmalignant tissue within the body of humanity.

Limits of God's Judgment

The extent of human defilement also sets the boundary lines of God's judgment. His wrath never goes beyond the extent of reprobation's damage. This principle comes into focus not only in the Genesis flood account but also in other portions of Scripture. One example appears in God's interaction with Abraham over the fate of Sodom.

As we read in Genesis 18:17–33, the Lord promised Abraham that if only ten righteous people could be found in Sodom and its sister city, he would save the entire populace for the sake of those ten. The Lord also promised that in destroying those cities, he would first rescue the one redeemable person and the relatives who chose to go with him. God's wrath never falls upon the righteous[22]—except in the case of Jesus, who willingly took our place.[23]

In a previous conversation, the Lord had told Abraham that the time had not yet come to deal with the Amorites living in the surrounding hills.[24] Their wickedness, the Lord said, had "not yet reached its full measure."[25] It had not developed to a degree that warranted their removal. God's judgment would be held back until the Amorites' wickedness reached totally reprobate proportions. This degree of degradation, Abraham was told, would take another 400 years.

Another example appears in God's instructions to the Israelites as they prepared to invade the land of Canaan under Joshua's leadership.[26] In the case of some Canaanite cities, the Israelites were instructed to kill all the people and their animals. In the case of other cities, God decreed death for the entire human population but not for their *nephesh* (soulish) animals.

The damaging impact of human evil on birds and mammals is something many of us have witnessed. *Nephesh* animals are highly motivated to please their human owners. If their human owners take pleasure in vicious behavior toward other people and animals, then that's the behavior they'll manifest. The book of Exodus instructs that if an ox has the habit of goring, the ox's owner must be informed of the behavior and ordered to correct it. If the ox later kills a human because the ox's owner has failed to restrain its behavior, then both the ox and the owner are to be put to death (Exodus 21:29). Exodus leaves no doubt that management of *nephesh* behavior is the responsibility of the animal's owner.

During the Israelites' conquest of Canaan, God rarely directed the Israelites to destroy everything: people, their soulish animals, and their material possessions as well. The extent of mandated destruction was determined by the extent of defilement. Such destruction resulted, naturally, in the death of insects, plants, etc., in the immediate area. Although these life-forms were largely unaffected by reprobate behavior, saving them was neither practical nor necessary. Unlike birds and mammals, these species multiply and reestablish themselves rapidly enough that a limited region of destruction would quickly recover.

In the rarest cases, such as in Sodom and Gomorrah, even the land (the region around the south shore of the Dead Sea)[27] was laid waste. To this day, despite the land's former fertility and abundant water supply, neither crops nor herds can be found there.

The account of the dialogue between Abraham and the Lord (Genesis 18:20–33) reveals a divine principle: God limits the scope of his acts of judgment. The extent of his "surgery" to remove societal reprobation is determined by the spiritual and geographical extent of that reprobation. This is a principle—and a promise—God has never broken and will not break. To do so would be a violation of his revealed character.

Application to Noah's Flood

This principle of God's precisely limited judgment holds implications for our understanding of Noah's flood. It means that if humans had spread as far as Antarctica and Australia, the flood would have submerged both territories, destroying all the people and animals impacted by human evil. However, if no one lived in Antarctica or Australia at that time, God would have no reason to submerge either continent or to kill any animals living there. Nor would Noah have been required to take a pair of emperor penguins or kangaroos aboard the ark.

The Genesis flood's extent, according to the principle and promise set forth in Scripture, would be determined by the spread of human habitation. If humanity had spread throughout Africa, Asia, and Europe, then those continents would have been submerged by the flood. If only Mesopotamia, the Persian Gulf area, and a portion of the Middle East had been settled, then only those regions would have been flooded.

Such a geographically bounded flood would still have been "worldwide" with respect to the human population and their domesticated animals, but not the entire globe. (See chapter 12 for biblical treatment of worldwide events.) A flood that exterminated all humans, all their soulish animals, and all their material possessions—outside Noah's ark—would be "universal" by definition and would achieve God's stated purpose in pouring out judgment. It would be sufficient to rescue Earth and its creatures from the defilement of reprobation. It would optimally express God's love and care for his creation.

Chapter 11

Dating Noah's Flood

To assign a precise date for the people and events of Genesis 6–9 is an impossible task. Nevertheless, many Bible interpreters over the years have attempted it. Virtually all these proposed dates fall within the recent range, some as recent as a few thousand years ago. However, these recent dates fail to withstand careful scrutiny, whether from a biblical or scientific standpoint.

In their book *The Genesis Flood*, John Whitcomb and Henry Morris claim that Noah's flood could have occurred as much as 3,000–5,000 years before the time of Abraham, who by their calculations was born in 2167 BC.[1] Although they allow for such a recent date, they argue likely gaps in the Genesis 5 and 11 genealogies imply a date for Noah's flood of about 5000 BC.[2]

Answers in Genesis (AIG), the world's largest and most widely known creationist organization, says the flood occurred on one of two possible dates: either 2348 BC or 2458 BC.[3] These dates differ just slightly because of a minor variance in the method by which they were derived. The first date came from simply adding up the years specified in the Genesis 5 genealogy—assuming no gaps in that listing. The second and more complex method used chronologies in Genesis, Exodus, Leviticus, Numbers, and Deuteronomy to determine a date for the exodus from Egypt, 1446 BC, which could be used to show that Abraham would have been born in about 2166 BC. From there, using the Genesis 11 genealogy to count backward in time from Abraham's birth—again assuming no gaps in the genealogy, and assuming the Genesis creation days to be six consecutive 24-hour days, yielded the 2458 BC date.

Chemist Jonathan Sarfati at Creation Ministries International calculated a date for Noah's flood, assuming no gaps in the Genesis 5 and 11 genealogies:

2522 BC.[4] Physicist Gerald Aardsma, formerly associated with the Institute for Creation Research but now working on his own, used biblical chronologies *and* the rise of the Dead Sea water level in his attempt to determine the flood's date. According to his calculations, Noah's flood occurred in 3520 BC.[5]

All current proponents of Noah's flood as a globally extensive event propose dates for it more recent than 5,000 BC. As noted in chapter 2, geologists who consider Noah's flood as a less-than-global historical event impacting some fraction of the human population propose flood dates ranging from 7,000–13,000 years ago. However, all these dates overlook multiple plausibility challenges from both biblical and scientific history. An abundance of indicators shows that Noah's flood occurred much earlier than 13,000 years ago. What follows are explanations of eight such indicators: four biblical and four historical/scientific.

Biblical Indicator #1: Incomplete Genealogies

In assigning recent dates for Noah's flood, global flood proponents incorrectly assume that the Genesis 5 and 11 genealogies are complete, or nearly so. That is, they presume that few, *if any*, generational names are missing from these lists. Meanwhile, a wealth of biblical scholarship confirms the existence of significant gaps—gaps of sufficient magnitude to allow for dating events such as Adam and Eve's creation and Noah's flood as occurring several tens of thousands of years ago, at least. This scholarship, in part, takes into account the fact that Hebrew words for father and son, *'ab* and *ben*, have a broad range of meaning.

'Ab refers to father, grandfather, great-grandfather, great-great-grandfather, and beyond.[6] It can be used for a forefather, a distant ancestor, or an individual possessing authority, such as a king, a prophet, or a priest.[7] Similarly, *ben* may mean son, grandson, great-grandson, great-great-grandson, and so on, or a later member of a tribal unit or social group.[8]

In the book of Daniel, we see that Belshazzar's mother refers to Nebuchadnezzar as Belshazzar's father (Daniel 5:11) when, in fact, two other kings came between them. What's more, Belshazzar and Nebuchadnezzar were not biologically related.[9] This flexibility in the usage of *'ab* and *ben* explains apparent discrepancies between or among parallel biblical genealogies, including those in 1 Chronicles 3, Matthew 1, and Luke 3. Similarly, the Hebrew verb, *yalad*, translated as "begat" 27 times each in Genesis 5 and 11 does not always refer to the birth of one's child. It can refer, for example, to a grandchild, as is the case in Genesis 46:15, 18, 22, and 25.

According to Exodus 6:18, Kohath had four sons: Amram, Izhar, Hebron, and Uzziel. Exodus 6:19 says the "sons" of Amram were Aaron and Moses. Numbers 3:27–28 records that while Moses was still living, some 8,600 male descendants were born to Amram, Izhar, Hebron, and Uzziel. This number would indicate that more than 2–4 generations separated Amram, Izhar, Hebron, and Uzziel.

Even the seemingly detailed genealogy in Genesis 11 must have omitted at least one name. The parallel record in Luke 3 includes Cainan, while Genesis 11 does not.[10] Furthermore, the Luke 3 genealogy mentions 11 individuals in the line from Shem to Abraham, while Genesis 11 mentions only 10.

Proponents of a flood date more recent than 3000 BC claim that some ancient copyist accidentally added Cainan into the Luke 3 genealogy.[11] However, only one known, not so early, New Testament manuscript, Codex D/Bezae (450–550 AD) is without the name Cainan.

Gaps in the Genesis genealogies are not to be considered flaws or errors, as some might think today. Genealogies are to be recognized as abbreviated records. The words translated into English as "When X had lived Y years, he became the father of Z" could just as accurately mean something slightly different to a scholar of biblical Hebrew. The latter would understand these words to say, "When X had lived Y years, he became the father of a family line that included, or culminated, in Z."

The problem with deriving a date for Noah's flood from the biblical genealogies lies in the wide degree to which these lists vary in their completeness. Comparative analysis of overlapping genealogies throughout the Bible suggests that they range anywhere from about 1 percent to 50 percent complete. If we were to estimate the timing of Noah's flood from biblical genealogies alone, applying the Genesis 11 time range between birth and giving birth of 29–100 years, we'd come up with a date between about 5,000 and 100,000 years ago.

Biblical Indicator #2: Division of the World

The Genesis 11 genealogy lists a total of 10 names: Shem (one of Noah's sons), Arphaxad, Shelah, Eber, Peleg, Reu, Serug, Nahor, Terah, and Abram. From both biblical and extrabiblical records, scholars have established that Abram (Abraham) lived approximately 4,000 years ago. Genesis 10:25 states that "the earth was divided" in the days of one of Abram's progenitors, Peleg, born several generations before Abraham. Given the larger context of the dispersal of humanity onto all Earth's landmasses (Genesis 11), this division of the earth most likely refers to the breakup of the great land bridges, a geologic process dating back to the end, or near the end, of the most recent ice age. Among the land bridges known to have been broken are the ones that joined Siberia to Alaska, Britain to France, Korea to Japan, and Haida Gwaii (Queen Charlotte Islands) to mainland British Columbia.

Scientific support for this understanding of Genesis 10–11 comes from recent research findings. Carbon-14 dating shows that relative sea levels rose rapidly, making both the Bering and Hecate Straits impassable to humans approximately 11,000 years ago.[12] This sea level rise was brought on by the melting of massive, thick continental ice sheets. In geographical regions where continental ice sheet melting occurred, some of the sea level increase was mitigated by isostatic rebound, the rise of landmass caused by reduction of ice weighing down the land. Stories recorded on totem poles crafted by the ancient people living in Haida Gwaii show that the surrounding sea level rose so rapidly that by 10,500 years ago, people could no longer emigrate easily to what is now British Columbia.[13] If Peleg lived roughly 11,000 years ago, and Abram about 4,000 years ago, and if the patriarchs' lifespans recorded in Genesis 11 are approximately proportional to the passage of time, then Noah would have been alive, very roughly, between 40,000 and 70,000 years ago.

Biblical Indicator #3: Early Date for Adam and Eve

An additional clue as to the time of Noah can be drawn from what the Bible says about the Garden of Eden, where Adam and Eve lived for at least a brief period. Genesis 2:10–14 names four rivers—Pishon, Gihon, Tigris, and Euphrates—that once flowed from three biblically specified sources: the mountains of Havilah in west central Arabia (the Pishon); the mountains of Cush in southwestern Arabia/eastern Africa (the Gihon); and the mountains north of the Mesopotamian plain (the Tigris and Euphrates).

The Tigris and Euphrates still flow today. Satellite imagery shows the dry riverbeds of two other major rivers that once flowed from the southwestern and west-central parts of the Arabian Peninsula, respectively, into the Persian Gulf region just south and east of Qatar (see figure 11.1). During the last ice age, melting snow and ice atop the mountains of southwestern and west central Arabia would have fed both rivers.

Genesis 2:10 says that the Pishon, Gihon, Tigris, and Euphrates Rivers all flowed near each other in the Garden of Eden. The only time the four rivers

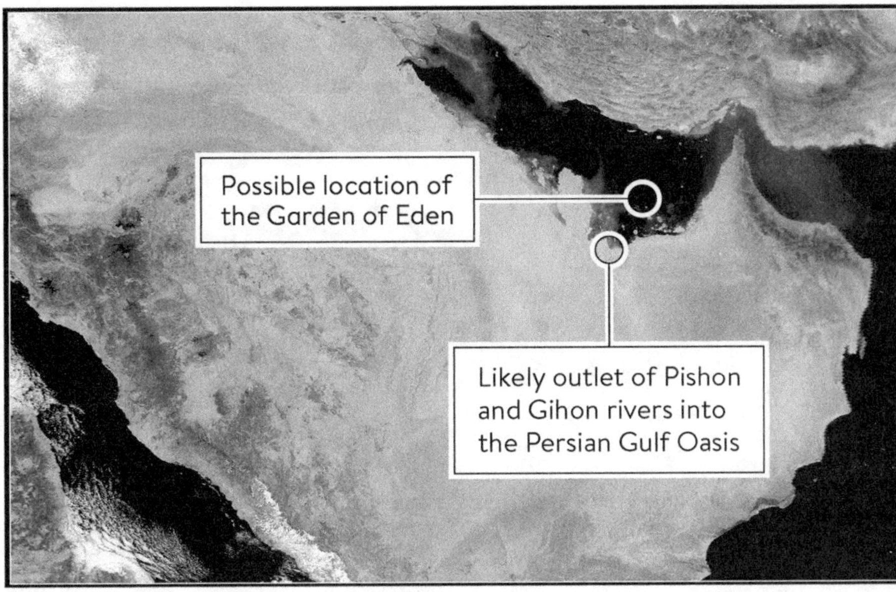

Figure 11.1: Satellite Image of the Arabian Peninsula
A high-resolution version of this image (available at visibleearth.nasa.gov/images/53880/saudi-arabia-and-dust-over-persian-gulf) shows the Tigris and Euphrates Rivers in Mesopotamia and two dry riverbeds originating in the mountains of the west central and southwestern parts of the Arabian Peninsula and descending into the Persian Gulf. *Credit: SeaWiFS Project, NASA/Goddard Space Flight Center, and ORBIMAGE*

would have been geographically and visibly close together is when most of the Persian Gulf was dry land. As it turns out, multiple times during the last ice age (which lasted from 120,000 years ago to 12,000 years ago[14]) a sizable portion of the Persian Gulf dried enough to form the "Persian Gulf Oasis." The location where the Pishon, Gihon, Tigris, and Euphrates would have flowed near to one another is, today, under the southeastern part of the Persian Gulf (see figure 11.1).

Had Noah's flood been globally extensive, transforming nearly all Earth's geological features and invoking catastrophic plate tectonics, then to identify preflood geographical features such as these four rivers and riverbeds, much less the possible location of the Garden of Eden, would be impossible.[15]

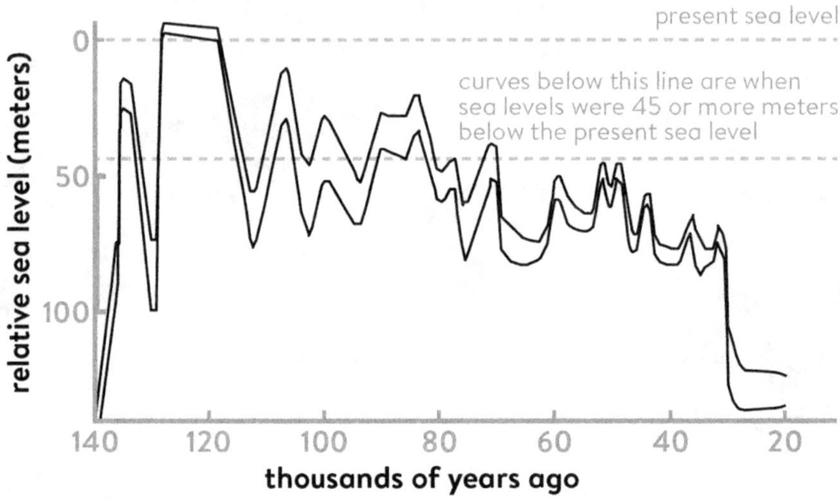

Figure 11.2: Sea Levels During the Last Ice Age Inferred from Coral Reefs
Credit: Data is from Kurt Lambeck and John Campbell, Science 292, no. 5517 (April 27, 2001): 680

Recent discoveries also lend support to belief in an early date for the creation of Adam and Eve and, thus, for Noah's flood. The average depth of the Persian Gulf is, at present, only 35 meters (115 feet). About 80% of the gulf today has a water depth of less than 45 meters (148 feet).[16] The most likely location for Eden sits, today, at a water depth of 40–45 meters (131–148 feet).

Only when Earth experienced one of the colder episodes within an ice age would so much of the Persian Gulf have been dry land. During such times, a large quantity of ocean water becomes locked up in ice sheets thousands of feet thick covering high-latitude landmasses.

Geologists Kurt Lambeck and John Chappell lend additional support to this early timing for the creation and flood by their measurement of the height and age of coral reefs in Huon Peninsula, Papua New Guinea (see figure 11.2) and of global ice volume (see figure 11.3). By two methods they determined that sea levels were significantly lower during the coldest years of the most recent ice age.[17] Another team, led by Tamara Pico, analyzed sedimentary cores in the Yellow River Delta to discover a similar record of global sea level history during the last ice age.[18] All three methods together reveal that throughout

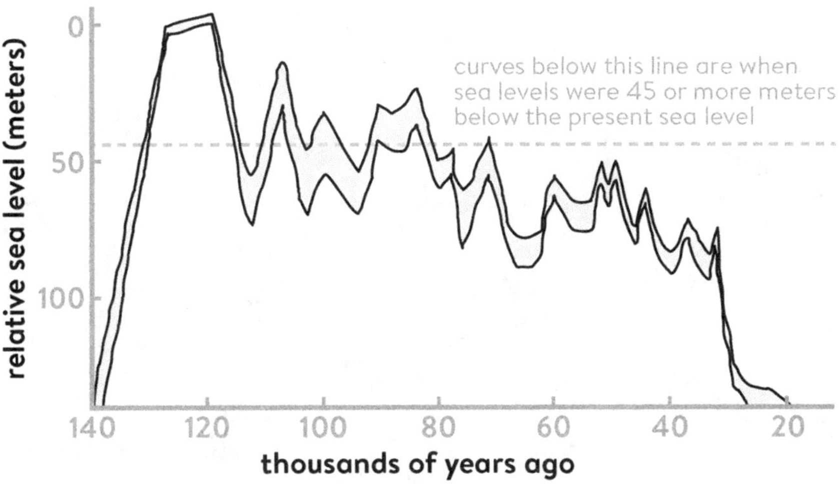

Figure 11.3: Sea Levels During the Last Ice Age Inferred from Global Ice Volumes
Credit: Data is from Kurt Lambeck and John Campbell, Science 292, no. 5517 (April 27, 2001): 682

most of the past 115,000 to 18,000 years, sea levels were more than 45 meters (148 feet) lower than they are today. However, Lambeck and Chappell caution that complicating factors, such as tectonic uplift of landmasses in response to weight reduction from the melting of thick ice sheets, can produce "considerable spatial variation."[19] That is, ice sheet dynamics cause different geographical regions to experience different sea level changes.

Biblical Indicator #4: Slow Recession of Floodwater

According to Genesis 8, the waters of Noah's flood took several months to fully recede. For the waters' recession to take that long, huge quantities of melting snow and ice must have replaced the water that would have flowed relatively rapidly into Earth's oceans. Even with plains as flat as the flattest parts of Mesopotamia and the Persian Gulf area, the rainwater and water from underground aquifers (Genesis 7:11–12) would have flowed out into the sea in less than a few months. So, for the floodwaters to remain for as long as Genesis 8 indicates, another source of water must have been contributing.

That source would most likely have been provided by a rapid and temporary

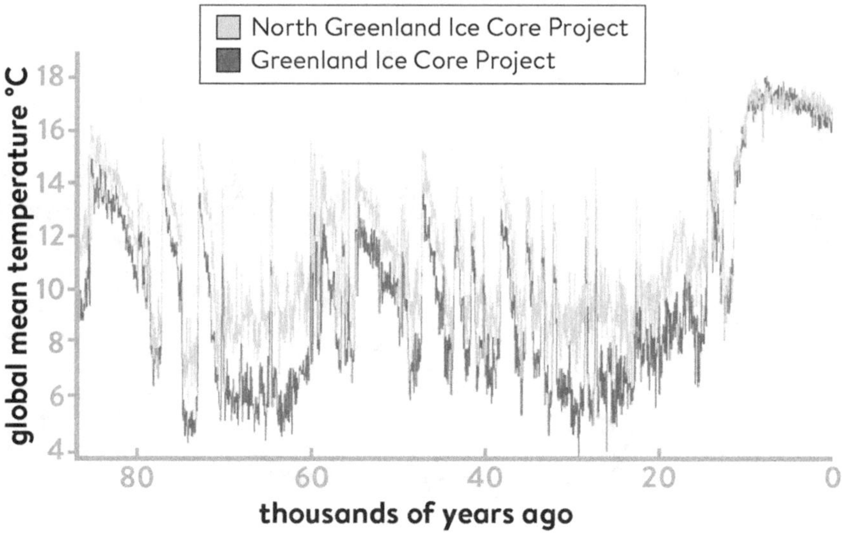

Figure 11.4: Rapid, Major Temperature Excursions During the Last Ice Age
Credit: light gray: North Greenland Ice Core Project; dark gray: Greenland Ice Core Project

warming event during the last ice age. This warming would have resulted, over a several-month period, in the melting of large quantities of snow and ice from surrounding high mountains. As figure 11.4 shows, at least a dozen brief and rapid warming events that occurred during the last ice age could have provided sufficient water to slow the flood's recession. These warming events occurred between 80,000 and 14,000 years ago.

Historical-Scientific Indicator #1: Lost Civilizations

Many ancient cultures retain stories of civilizations, cities, and villages that became lost as they sank below sea level. The best known, Atlantis, was mentioned by Plato in his written dialogues, *Timaeus* and *Critias*. The best scientifically documented example comes from archaeological studies along the Persian Gulf shoreline.

In the journal *Current Anthropology*, archaeologist Jeffrey Rose reported on research conducted over six years that revealed the remains of more than 60 previously unknown ancient village sites along the shoreline of the Persian

Gulf.[20] Each of these sites was dated at 7,500 years old, at least, and possibly much older. In a follow-up interview by Live Science writer Jeanna Bryner, Rose noted, "these settlements boast well-built permanent stone houses, long-distance trade networks, elaborately decorated pottery, domesticated animals, and even evidence for one of the oldest boats in the world."[21]

Archaeologists Hans-Peter Uerpmann and Margarethe Uerpmann, in their studies of the remnants of three ancient villages at the base of Jebel Faya (a limestone mountain on the south shore of the Persian Gulf), concluded that people in that region had relocated in response to a sea level rise in the Persian Gulf.[22] As the sea level rose, they abandoned their drowning shoreline villages and built new ones at a slightly higher elevation. Optical luminescence dating, an indirect dating method with large systematic uncertainties (see appendix), indicated that the three villages were constructed at least a few tens of thousands of years ago.[23]

Rose proposed that during the latter part of the last ice age, a thriving civilization existed on land that is now below the current level of the Persian Gulf. He deduced that as the sea level rose, people fled the Gulf Oasis and built a succession of settlements along the Persian Gulf's rising shoreline. Rose suggested that underwater exploration of the Persian Gulf may yield confirming evidence for his theory. Even apart from this additional research, he affirmed the geological evidence for the rapid and chaotic rise of the Persian Gulf's sea level during the latter part of the last ice age. This rapid, chaotic sea level rise, he noted, most likely destroyed any evidence of human settlements that may have existed at that time.

Archaeological evidence provided by Rose and the Uerpmanns, along with what the Bible says about Eden's location, the slow recession of the floodwaters, and "the dividing of the world in the days of Peleg" together imply that Noah's flood occurred sometime before the latter portion of the last ice age. That is, it must have occurred well before 15,000 years ago, at least.

Historical-Scientific Indicator #2: Emergence of Linguistic Diversity

Genesis 11:7–9 describes a time *after* Noah's flood, by several generations, when human language diversified from one to many. Thus, if linguists can establish when the diversification of human language occurred, they will provide an important clue as to when Noah's flood occurred. We can be confident, from the biblical text, that the flood predated that time.

From written literature, oral stories, archaeological relics, and cultural clues from the past 10,000 years, linguists seek to determine the rate at which various languages developed throughout human history. This method, however, is extraordinarily difficult to apply, and the dates derived are only approximate. As *The Handbook of Historical Linguistics* reminds us, "Language change is not clock-like. Languages can vary substantially in their rates of lexical replacement."[24]

With this caveat noted, linguistic research has provided at least an approximate date when one of our major language groups diversified from a single language. Studies indicate that the family of Indo-European languages diversified from a single language before 8,000–10,000 years ago.[25] Transeurasian languages began to diversify before 9,000 years ago.[26] An analysis of 400 Austronesian languages suggests they originated from a single language 13,000–17,000 years ago.[27] This latter date implies that Noah's flood occurred sometime earlier than 13,000–17,000 years ago.

Historical-Scientific Indicator #3: Humanity's Spread

In Genesis 9:1 God repeats to Noah and his descendants the command originally given to Adam and Eve: "Multiply and fill the earth." God repeats the command yet again in Genesis 9:7, instructing Noah's descendants to "be fruitful and multiply; spread out over the earth and multiply upon it." However, in Genesis 11:4 we read that postflood humans had resolved to remain in one region and firmly resisted spreading to all Earth's landmasses. According to the text, they made this bold plan: "Let us build ourselves a city and a tower with its top in the sky. Let us make a name for ourselves; otherwise, we will be scattered throughout the earth." Genesis 11:7–9 tells the story of God's response to this rebellion, a stubborn refusal to disperse beyond their familiar territory. He intervenes in a dramatic way to scatter humanity throughout the earth.

No precise method has been developed to date when this scattering began. However, scientists can provide a reliable range of dates through analysis of human fossil remains, archaeological artifacts, evidence for agriculture and food preparation, and genetic clues. Using these four tools, scientists have been able to determine when humans first migrated to and settled in Europe, eastern Asia, Australia, and the Americas.

The earliest undisputed evidence for the remains of anatomically modern humans in Europe dates back to 45,000–39,000 years ago.[28] The earliest undisputed archaeological artifacts associated with these anatomically modern humans date to 42,000–38,000 years ago.[29] The earliest evidence in Europe for the harvesting, roasting, and grinding of grains to make bakery products dates to 32,600 years ago.[30] Carbon-14 dating shows that food preparation, as

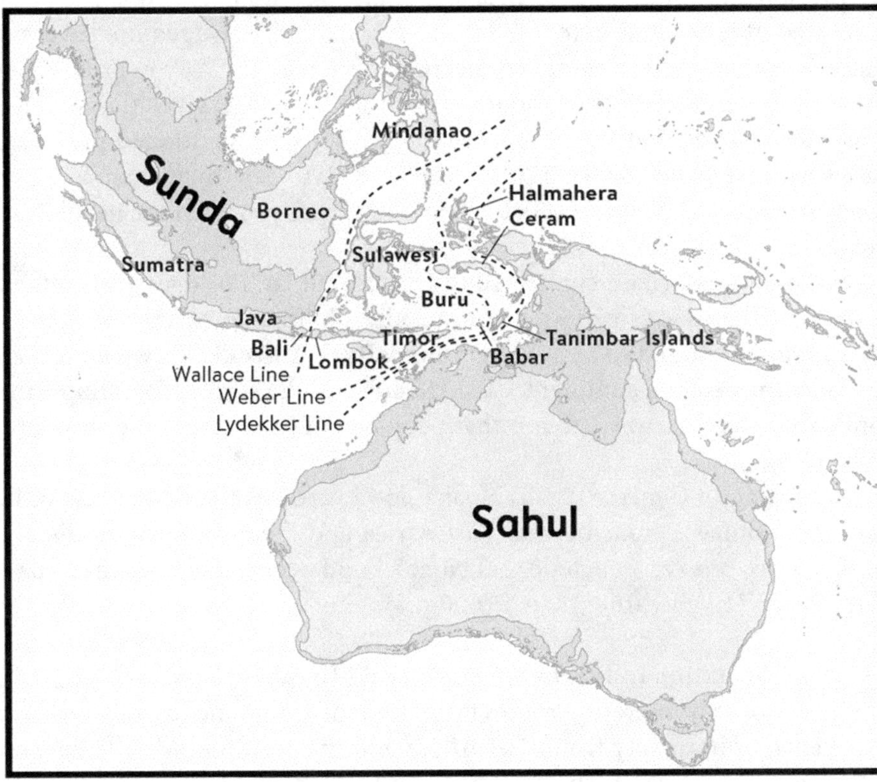

Figure 11.5: Sunda and Sahul During the Last Ice Age
Credit: Kanguole, Creative Commons Attribution–Share Alike

evidenced by soaking, grinding, and charring of seeds, occurred in Greece as early as 40,000 years ago.[31] Reviews of multiple independent genetic studies indicate that humans lived in Europe as early as 45,000–39,000 years ago.[32] These consistent results strongly indicate that humans began to settle in *both* northern and southern Europe 45,000–40,000 years ago.[33]

The same four dating methods establish that anatomically modern humans began to occupy virtually all of Asia's land area by as early as 50,000–40,000 years ago.[34] The first human settlements in Australia date back to 47,000–40,000 years ago.[35] For North and South America, the estimated date is 16,500 years ago.[36]

The spread of humanity to nearly all Earth's continents and islands prior to the Neolithic revolution (c. 11,000 years ago) would have required access to relatively easy migration routes connecting these regions. We can be confident that such routes existed throughout most of the last ice age, given that sea levels were lower during that time. On several occasions, as mentioned previously, sea levels were lower by 45–120 meters (148–390 feet) than their levels are today (see figures 11.2 and 11.3). Such low sea levels provided the needed land bridges.

Not only was Britain connected to France, but land bridges also connected most of the islands of Indonesia and the Philippines to the mainland of Asia, a region referred to as Sunda or Sundaland. Australia, New Guinea, and Tasmania formed one continent, called Sahul, and the straits separating Sunda from Sahul were narrow enough that people could see across the straits (see figure 11.5).

According to Genesis 10–11, Noah's flood predates the first efforts of humans to colonize regions beyond East Africa and the Near and Middle East. On this basis, one can conclude that Noah's flood occurred earlier than 40,000 years ago, and likely earlier than 45,000 years ago.

Historical-Scientific Indicator #4: Postflood Migration Routes

Emergence of easy migration routes facilitated the simultaneous, relatively rapid dispersal of humanity from East Africa and the Near and Middle East into all the regions of Asia, Australia, Africa, and Europe. Land bridges connected the islands of Indonesia and the Philippines to the Asian mainland, Australia to New Guinea, Britain to France, southern Arabia to East Africa, Asia Minor to Greece and Cypress, Iberia to North Africa, Britain to Ireland, Taiwan to mainland China, the islands of Japan to the Chinese and Russian mainlands, Norway to Germany, India to Sri Lanka, and Italy to Corsica, Sardinia, Sicily, Tunisia, and Croatia (see figure 11.6). These land bridges, which at the time were optimal for facilitating human migration, made possible the nearly simultaneous settlement of virtually all of Africa, Asia, Australia, and Europe.

At the time of humanity's dispersal, described in Genesis 10–11, land bridges also connected eastern Siberia to Alaska and Greenland and Baffin Island to the North American mainland. However, these land bridges were covered with far too much ice to be passable by humans until about 16,500 years ago. This limitation is not an issue for the account of humanity's dispersal described in Genesis 10–11. None of the people groups and geographical locations mentioned in Genesis 10–11 pertain to North or South America.

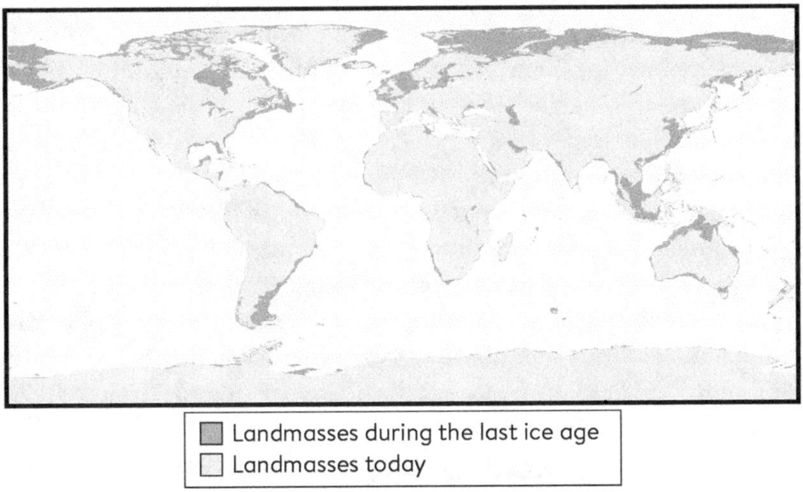

Figure 11.6: Global Sea Levels During the Last Ice Age
Credit: National Oceanic and Atmospheric Administration

While the land bridges that existed previous to 40,000 years ago enabled humans to migrate into virtually all of Africa, Asia, Australia, and Europe, coastal routes greatly sped up the migrations and accelerated the settlements. Beaches along the shores of continents and islands made for relatively easy walking and transportation of goods. Small boats and rafts enabled people to get around coastal cliffs and estuaries.

Coastal environments typically are endowed with abundant and reliable food sources (for example, fish, shellfish, seaweed, shoreline vegetables and fruits, rodents, and birds) and fuel (driftwood, tree sap, dried seaweed, and bushes). Temperature variations in coastal environments are less extreme than in inland regions. Slopes leading down to coastal beaches provide fresh water from streams. Rivers emptying into the sea provide easy access into the deep interior of continents and islands. The role of coastal routes that made possible the rapid, simultaneous migration into and settlement of all the regions of Africa, Asia, Australia, and Europe and the later migration into and settlement of North and South America is well documented and established in the scientific literature.[37]

The advantages of seashore environments became clear to me when I was just ten years old. My dad took my sisters and me to a deserted beach along the coast of British Columbia. Amid rainy conditions, he showed us how to build a small lean-to from available brush and collect dry tinder from driftwood cores. After starting a small fire under the small lean-to, we went to work on building a larger lean-to, near the fire, for ourselves. Then we gathered clams and oysters off the beach and picked wild berries, onions, and lilies from shoreline bushes for some tasty meals. That day we saw how much easier it can be to survive and travel along an ocean shoreline than through a thick interior forest.

Archaeological, fossil, and genetic evidence for widespread dispersal of humanity 50,000–40,000 years ago shows that migrants took advantage of river deltas and other shoreline locales. Many chose to settle where resources were abundant. Others would move on. This migration pattern led to a rapid growth in human population. God's command to Noah and his family to "be fruitful and increase in number; multiply on the earth and increase upon it" (Genesis 9:7) was finally being fulfilled.

What today seems the biggest barrier to migration between the Mesopotamia-Persian Gulf region and eastern Africa is Rub' al Khali (the "Empty Quarter"). This area, which encompasses the southern third of the Arabian Peninsula, is nothing but sand, gravel, and gypsum. Its average daily maximum temperature throughout the entire year is 47°C (117°F)!

A study of alluvial fan deposits in Rub' al Khali has revealed that during the last ice age, the region experienced greatly increased rainfall (yielding extensive vegetation) on two occasions, roughly 55,000 years ago and 75,000–120,000 years ago.[38] The study showed the development of sufficient precipitation to activate a drainage system in Rub' al Khali, a system that corresponded with increased monsoon activity. In addition to these episodes, deglaciations during the last ice age stimulated by intermittent increases in Earth's orbital eccentricity would have brought periods of humidity to Rub' al Khali.

The availability of fresh water and vegetation in Rub' al Khali would have made human migration across the land bridge connecting eastern Africa to southwestern Arabia easy to navigate, even inevitable. Genesis 2:13 describes the Gihon River flowing out from the mountains of Cush into the Garden of Eden. Cush is normally identified with the mountainous region within and surrounding eastern Ethiopia. However, during the last ice age, that mountainous region would have included the mountains in the southwest tip of the present-day Arabian Peninsula. Satellite imagery reveals a dry riverbed extending from that southwest tip into the southern part of the Persian Gulf. That river, flowing with water and with abundant plants growing along its banks, would have presented an inviting, relatively cool transportation corridor connecting the Persian Gulf region to eastern Africa.[39]

Paleolake deposits provide evidence for a second freshwater corridor connecting the Persian Gulf region to the Arabian interior as well as to the Levant (geographical region bordering the Eastern Mediterranean between present-day Turkey and the Sinai Peninsula) and eastern Africa.[40] Several times between 130,000 and 80,000 years ago such a transportation corridor existed.

An interdisciplinary research team discovered human footprints in an ancient lake deposit in southwestern Arabia with no accompanying archaeological evidence—indicative that the lake was a mere stopping point along a migration route. Optical luminescence dates for two sets of prints were 121 ± 11 and 112 ± 10 thousand years ago.[41] (Note that error estimates for the two dates did not include systematic errors, which could be much larger than the statistical errors [see appendix]).

A team of paleontologists and ecologists affirmed that this easy migration route connecting the Middle East to Africa was accessible at several epochs during the last ice age.[42] The team found fossils of an extinct subspecies of the eastern crested rat, *Lophiomys imhausi maremortum*, in the Levant. Previously, fossils of *Lophiomys imhausi maremortum* had been found only in central-eastern Africa.

The discovered fossils date from 103,000 to 42,000 years ago. This discovery establishes that an easy migration route connecting central-eastern Africa and the Levant existed at various times between 103,000 and 42,000 years ago. If rats were able to migrate from central-eastern Africa to the Levant, it must have been possible for humans at that time to travel back and forth between the two regions. Thus, at least two accessible migration routes connected the Middle East with Africa at times during the last ice age. Their availability would have facilitated the major dispersals of humanity described in Genesis 10–11. The fact that easy migration routes connecting the Persian Gulf region, Arabia, Mesopotamia, the Levant, and eastern Africa persisted as far back as 130,000 years ago suggests that Noah's flood likely extended to a much wider region than just the Persian Gulf Oasis and Mesopotamia.

Too Cold?

Global flood proponents argue that ice age temperatures would have been too cold to allow for construction of the ark or for rapid reinvigoration of agriculture after the floodwaters had receded. If Noah had been living in a high latitude region, cold would have been a problem, but he did not. He likely was living in the plains of the Persian Gulf Oasis, the part of the present Persian Gulf that is dry land during an ice age.

When we think about the warmest places on Earth, our minds go to the tropics. Southern India, Borneo, New Guinea, the Congo, and South Sudan are well known for their extreme heat. As for marine regions, the South China Sea, the Coral Sea, and the Java Sea are equally recognized for their warmth. The world's warmest sea, however, is not in the tropics. It's north of the tropics. The Persian Gulf holds the high-temperature record for Earth's seas.[43] The southern part of the Gulf is significantly warmer than the mid to northern parts.[44] During the last ice age, the southern portion of the Persian Gulf Oasis would have been as far away from thick ice sheets as Edmonton, Alberta, is from Hermosillo, Mexico.

With sea levels 45–120 meters (148–390 feet) lower than today's level at several times during the last ice age, nearly all of what is now the Persian Gulf would have been dry land. Temperatures rise by 1.9°C (3.5°F) with every 300-meter (1,000-foot) decline in elevation. As figure 11.4 shows, there were several episodes, each at least three centuries long, during the last ice age when global mean temperatures were only 2–3°C colder than they are today. Thus, there would have been times during the last ice age when Noah would have sufficiently moderate temperatures for a long enough time to build the ark before the flood and to reinvigorate agriculture after the flood.

Estimated Date for Noah's Flood

In the date range of 250,000–45,000 years ago, scientists lack a useful radioisotope decay method for dating human remains and artifacts. In this date range, they have no recourse but to use indirect dating methods that involve large known and unknown systematic errors. (Systematic errors refer to environmental and/or instrumental factors that could shift all measurements up or down in value.)

The systematic errors can be substantial. For example, the long-held claim—based on thermoluminescence dating of circular engravings in the Jinmium Rock Shelter—that modern humans inhabited Australia 75,000–58,000 years ago[45] was shown to be incorrect by a factor of twenty times. Carbon-14 measurements of charcoal samples found in the same layer as these engravings yielded dates that ranged from 3,900 to 100 years ago.[46]

The systematic errors in the indirect dating methods applied to human remains and artifacts almost always shift dates to older values. Thus, dates obtained by these indirect methods in nearly all cases should be interpreted as upper limits on the time back to the sample's actual date. (For explanations of these systematic errors and their expected magnitudes, see the appendix.)

In light of the relevant carbon-14 dates and the indirect dating methods where systematic errors *are* known, Noah's flood must have occurred previous to 40,000 years ago and very likely earlier than 45,000 years ago. Given the biblical evidence that Noah's flood occurred during an ice age (see chapters 12 and 13) and the biblical implication that it probably occurred during the most recent ice age, it seems unlikely that Noah's flood happened any earlier than 120,000 years ago. As explained in chapter 13, a case can be made for Noah's flood taking place during one of the major, rapid melt events of the previous ice age. At least a dozen such melt events occurred between 80,000 and 45,000 years ago. Sometime within this date range would be a reasonable estimate of the flood's timing. A more conservative date for Noah's flood would be sometime between 120,000 and 45,000 years ago.

The following three chapters address frequently asked questions as to why God didn't instruct Noah simply to move his family and their animals to an uninhabited part of Earth? Why go to all the trouble of building such a huge ship? Chapters 12 and 13 also address the extent of Noah's flood from a biblical and a scientific perspective, respectively. Chapter 14 discusses the necessity and feasibility of the ark.

Chapter 12

Biblical Clues to the Flood's Extent

For some Christians, belief that the flood covered the entire planet and shaped nearly all its geological features has become a litmus test for biblical orthodoxy, essential evidence of belief that the Bible is God's true and trustworthy Word. For other Christians, the only way to defend the doctrine of biblical inerrancy is to say that Genesis 1–11 is a metaphorical, allegorical, or mythical story—*not* a historical or chronological account. Both groups consider their view as the key to linking Genesis with the gospel message. Each sees the other's view as invoking contradictions between what the scientific record clearly reveals and what the Bible teaches (see chapter 8).

The Case for Global Extent

Global flood proponents' supporting argument springs primarily from assumptions about the meaning of "world" and "earth." The rest of their case flows from this starting point. The wording of Genesis 7:19–21 *in English* may explain why so many English language readers of the Bible conclude that Noah's flood must have covered the entire planet. English translations of this passage strongly suggest that the floodwaters covered all the mountains "under the whole heaven." In the King James Version, the text says this:

> The waters prevailed exceedingly upon the earth; and all the high hills, that were under the whole heaven, were covered. Fifteen cubits upward did the waters prevail; and the mountains were covered. And all flesh died that moved upon the earth.

The New International Version says this:

> They [the floodwaters] rose greatly on the earth, and all the high mountains under the entire heavens were covered. The waters rose and covered the mountains to a depth of more than fifteen cubits. Every living thing that moved on land perished.

The text would appear to say that all land life on the planet was destroyed and that even Earth's highest mountains were covered by more than 20 feet of water.

While global flood proponents acknowledge that the biblical Hebrew words *'eres* and *'adama*, for "earth" or "world," used throughout Genesis 6–8 can be used with reference to a portion of Earth's surface, they argue that if God intended to communicate a less-than-globally-extensive flood, he would have inspired the author to use any of three other Hebrew nouns for earth. These other words, *haraba*, *yabbasha*, and *sayon*, never refer to the whole Earth.

A Matter of Context

You may be old enough to remember these lyrics: "He's got the whole world in his hands." When most of us first heard or sang these words, we pictured the entire planet and all its people. If we were living two thousand years ago or even just several centuries ago, the words would have meant something different.

Within the last few hundred years, nearly every human has developed a global perspective. Today a person with a robust travel budget can visit nearly any location on the planet. Even those who can't afford to travel to distant lands possess the technological means to see or hear about far-off places and to know at least something about those places. Virtually everyone on Earth today has access to cell phones, the internet, books, and world news. Most people have some awareness of happenings in other parts of the planet.

Today's global perspective inevitably colors our interpretation of Scripture. When we encounter such phrases in Genesis 7 as "under the entire heavens" and "every living thing on the face of the earth," we picture the whole surface of the globe. We think of the entire planet within a larger system of planets, just one among several thousand known systems of planets. This is not what people of earlier times pictured. What did "the entire heavens" and "the face of the earth" mean to the ancient Hebrews and their contemporaries? We must consider their frame of reference, not ours. Is there a biblical precedent for interpreting these words in a less-than-global way?

Nonglobal "Worldwide" Events in Scripture

Repeated references in Genesis 7–8 to Noah's flood impacting "the earth," "the face of the earth," "all the surface of the earth," and "throughout the earth" have caused many people in the past century or two to conclude Noah's flood must have been globally extensive. However, the Bible mentions several other "worldwide" events, in addition to Noah's flood, that were less than globally extensive. The first example is Joseph's feeding of the whole world. Genesis 41:57 says, "All the countries came to Egypt to buy grain from Joseph, because the famine was severe in all the world." Genesis 42:5–6 defines all the world of Joseph as limited to the provinces of the Egyptian empire, plus the land of Canaan.

The second is found in the account of foreign dignitaries' coming to receive King Solomon's wisdom. 1 Kings 4:34 says, "Men of all nations came to listen to Solomon's wisdom, sent by all the kings of the world." 1 Kings 10:24 adds, "The whole world sought audience with Solomon to hear the wisdom God had put in his heart." However, we read in 1 Kings 4:31 and 2 Chronicles 9 that these visitors came from as far away as Sheba (modern-day Ethiopia) and all the lands of Arabia. It appears the "whole world" of Solomon extended no more than about 2,100 kilometers (1,300 miles) from Jerusalem.

The New Testament also speaks of "worldwide," but less-than-global, events. For example, at the time of Christ's birth, "a decree went out from Caesar Augustus that all the world should be registered" (Luke 2:1, ESV). History records that Augustus's authority extended only as far as the geographical limits of the Roman Empire, not beyond.

At the first celebration of Pentecost following Jesus's resurrection, "there were staying in Jerusalem God-fearing Jews from every nation under heaven" (Acts 2:5). Genetic studies attest that none of the first-century Jews could have come from Bolivia or Australia.[1] For Jewish people living in first-century Jerusalem, "every nation under heaven" included the provinces of the Roman and Parthian Empires.

The apostle Paul twice refers to the Roman world as "all the world." In his letter to the Christians in Rome he writes, "Your faith is being reported all over the world."[2] Here, the world to which Paul refers is the vast, but not global, Roman Empire. Obviously, Paul did not mean the Inuit of the Yukon or the Polynesian Islanders had received news of the Christians in Rome. In his letter to the Colossian believers, Paul refers to the Roman Empire's extent—the world known to the church in Colossae when he writes: "All over the world this gospel is producing fruit and growing, just as it has been doing among you."[3]

Each of these biblical references to a "worldwide" occurrence points to a region much less extensive than Earth's entire land area. Based on these clearly established examples, phrases such as "the entire heavens" and "the face of the earth" that appear in Genesis in describing Noah's flood, may also be faithfully interpreted, in context, as an area much less extensive than the entirety of the globe.

"All," "Every," and "Everything"

The words "all," "every," and "everything" appear more than 40 times in Genesis 6–8 in reference to humans, animals, and mountains/hills. For many readers—and a few Christian scholars—this repetition leaves no possible meaning other than the entirety of Earth's surface.[4] They wonder what else could it mean.

As with any descriptive word, the meaning of "all," "every," or "everything" is determined by context. Context is essential for understanding the intended message.

The repetition of the words "all" and "every" in Genesis 6–8 does have interpretive significance. However, rather than referring to all Earth's continents and islands, this repetition more likely emphasizes the totality of the flood's impact on all people and their animals living at Noah's time (except those in the ark). If humans had not yet begun to occupy all of Earth's lands, Noah's flood could be worldwide, or "universal," *without* covering the entire planetary surface.

Which Animals Were Rescued?

Most English translations of Genesis 6–8 seem to give readers the impression that *all* species of land animals on Earth were impacted by Noah's flood. This impression raises both theological and scientific concerns of major significance not only to twenty-first century readers, but also to readers living in earlier centuries of Christianity and Judaism.

The theological problems arise from the notion that God destroyed vast numbers of animal species that had never been touched by the wickedness of Noah's contemporaries. Such an act would represent a violation of God's revealed character and counter to all other biblical examples of God's judgment. Divine action to cleanse the world of societal reprobation never extends beyond the scope of that reprobation (see chapter 10). Therefore, the devastation of Noah's flood came only upon animals, specifically *nephesh* (soulish) animals, irreparably damaged by contact with human reprobation.

The scientific problem is that even with the longer cubit measure, the ark is much too small to accommodate all species of land animals and all the provisions needed to keep them alive for a whole year. Furthermore, to care for all these animals would require more people than just the eight on board the ark.

Help in identifying the creatures lost due to Noah's flood—as well as those rescued in the ark—comes from three sources: first, from the brief time Noah was given to take animals on board the ark; second, from biblical teaching about the limits of God's judgment; third, from consideration of the Hebrew nouns used for animals in the Genesis account of Noah's flood.

Genesis 7:4–10 states that the time between God's instruction to take pairs of animals on board the ark and housing them was a mere seven days. A period this brief would limit the geographical extent from which these animals could be gathered. Antarctica, Australia, North America, Oceania, and South America would surely have been ruled out.

Speculation that years before the seven days, God may have motivated animal pairs to migrate thousands of miles from their habitats to the vicinity of Noah's ark runs into several problems. First, the Bible makes no mention of such migration. Second, such migration would require certain animal pairs to traverse regions devoid of their food sources or to encounter terrain impossible to cross. Third, some of the animal pairs lacked the necessary anatomy for such a journey. Examples include penguins, sloths, koalas, platypuses, and moles.

God always preserves and protects whomever and whatever can be rescued. This principle is consistent with the Hebrew words used in Genesis 6–9 to describe the animals spared from the flood. The text includes these seven nouns, defined in the *Theological Wordbook of the Old Testament*:

- ***basar***: flesh; animal musculature; mankind; animals used in Jewish sacrifices[5]
- ***behema***: beast; cattle; long-legged, four-footed mammal[6]
- ***hayya***: living animal; wild, not domestic, animal[7]
- ***nephesh***: soulish creature; person; mind; land creature with the breath of life; creature capable of expressing yearnings, emotions, passions, and will; self-aware creature[8]
- ***op***: bird; fowl; insect; winged creature[9]
- ***remes***: creeping living creature; short-legged land mammal such as a rodent; small reptile[10]
- ***sippor***: bird; little bird[11]

All these words refer to certain varieties of birds and mammals, though some can be used more broadly. This list of creatures compares closely with the list of animals God created on the fifth and sixth creation days of Genesis 1, animals significant to preparing Earth for humanity and, according to Job 38–39, crucial for launching and sustaining human civilization.[12]

Clearly, the survival of these creatures would be important to the restoration and progress of human society after Noah's flood receded. Nothing in the Genesis text compels us to conclude that Noah's passengers included anything other than the birds, mammals, and perhaps a few reptilian species that lived in the area then inhabited by humans.

Another insight into the nature of the animals on the ark is that they came willingly (Genesis 7:8–9) and quickly to Noah. This behavior gives another indication that the animals were capable of bonding to humans.

The bottom line is that the only animals on Noah's ark were *nephesh*, those having contact only with Noah and his family, animals not having suffered irreparable abuse by reprobate humans. The only animals earmarked for destruction by Noah's flood were *nephesh* animals irreparably damaged by reprobate humans.

What About the Flood's Depth, the Water's Height?

Most English translations of Genesis 7:19–20 would appear to say not only that all land-dwelling life on Earth was destroyed, but also that Earth's highest mountains and mountain ranges were covered by more than 20 feet of water for at least five months. Here, again, we must examine what the original Hebrew words literally convey. A careful look at these words paints a far less geographically expansive picture.

The Hebrew verb translated "covered" in Genesis 7:19 is *kasa*. This "covering" can be defined in any of four ways: residing upon, running over, falling upon, or concealing. Theologian R. Laird Harris in the *Theological Wordbook of the Old Testament* comments on the meaning of *kasa*:

> The usual usage of the verb *kasa* is the literal meaning "to cover." Frogs covered Egypt (Ex. 8:6). The pillar of cloud covered the tabernacle (Num. 9:16). It is also used more generally to mean "conceal" (Gen. 37:26, Prov. 10:18, KJV "hide") or "overwhelm" (Prov. 10:6, 11, NIV "overwhelm"). In Gen. 7:19–20 the hills were "covered;" the Hebrew words do not specify with what. The NIV specification of water goes beyond the Hebrew. The Hebrew may merely mean that the mountains

> were concealed from view by some kind of extreme meteorological event.[13]

The only certain conclusion one can draw from Genesis 7:19–20 is that from one horizon to the opposite horizon, all Noah could see from his perspective on the ark was water and clouds.

Whether the hiding or concealment of the hills was by water, these differences in the meaning of *kasa* are important. *Kasa* could be interpreted to mean that more than 20 feet of water stood, that is, remained, over the high hills or mountains. Or it could mean that this quantity of water ran over them, as in a flash flood, or fell upon them as rainfall. The context of Genesis 7 gives no indication which of the meanings is intended. Not that the choice is significant for understanding the effects of such "covering." Any of the three scenarios would guarantee widespread destruction, with no possible survivors.

The Hebrew words translated "all the high mountains," *kol heharim hugebohim*, are also open to a wide range of translation alternatives. Given that biblical Hebrew is limited to a vocabulary of about 3,000 words (not counting the names of people and towns), many of these words, especially nouns, have multiple distinct literal definitions. (For comparison, English includes a few million words and continues to grow by several thousand words per year, according to digital analysis.[14]) No wonder the most heated controversies over the meaning of Old Testament texts occur among English-language speakers, with Japanese and Korean speakers not far behind. (The vocabulary size of both modern Japanese and Korean exceeds 500,000 words.)

The word translated "mountains" in Genesis 7 is the same word used often in Scripture for "hills." *Har* can refer to either "hill," as in "hill country," or "mountain."[15] It can be used with reference to towering peaks that mountaineers love to ascend or to a small hill that children enjoy climbing in their playtime. Any landform between these two extremes fits within the range of meaning for *har*.

The Hebrew adjective *gaboah* also carries a range of meaning: "high," "exalted," "elevated," or "lifted above."[16] It applies to any elevation above the plains, from a landmark hill to a peak such as Mount Ararat. Genesis 7:19 depicts Noah's inability to see anything but water and clouds from his vantage point on the ark's upper deck, looking from one horizon to the opposite horizon. If the ark were floating anywhere near the middle of the Persian Gulf or over the vast Mesopotamian plains on water as much as 200 or 300 feet deep, no hills or mountains would have been visible to Noah.

From any such location at the height of the flood, Noah would see only water and clouds. The high mountain ranges surrounding the Mesopotamian region and the Persian Gulf would have been beyond Noah's line of sight—his view limited by Earth's curvature and by atmospheric conditions, among other factors. Those who travel through California's San Joaquin Valley (much narrower than the Mesopotamian valley) can rarely see the towering peaks of the Sierra Nevada Mountains to the east or the coastal range to the west—even more rarely, if ever, from the valley's center or in cloudy/rainy weather.

Further indication that Noah's flood covered a limited region of Earth—the humanly inhabited area rather than the whole globe—comes from Genesis 8:5, which says,

> The waters continued to recede until the tenth month, and on the first day of the tenth month, the tops of the mountains [*har* in its plural form] became visible.

Although Noah could see land by this time, neither the raven nor the dove he released could fly far enough to reach a landing place. After another week, the raven did not return, and Noah sent out the dove for a second time. This time it recovered a leaf from an olive tree. This is an important clue, given that we know where olive trees grow. They do not grow on high mountains. We can reasonably assume from this detail that the hills where the ark had come to rest and where the dove was able to find an olive leaf were low-lying hills or foothills in the vast range associated with Ararat.

Genesis 8:9 describes the dove's first flight. From the perspective of the dove as it flew, "the waters were still on the face of the whole earth" (ESV). Here we see a clear indication that "the face of the whole earth," in the context of Genesis 6–8, can and does mean an area less extensive than Earth's entire surface. The similarity of wording in Genesis 7:19 and 8:9 affirms that we need not interpret the account of Noah as saying that the flood covered all of Earth and all its mountains.

Words Used for Earth

As noted previously, two biblical Hebrew words commonly used for "earth" or "land" are *'eres* and *'adama*. *'Eres* can refer to the entirety of Earth's surface, all Earth's landmasses, or a fraction of Earth's land. *'Adama* has a more restricted meaning, as we see in Genesis 1. God made the first man, Adam, from the *'adama* to till the *'adama*. As theologian Leonard Coppes explains in the *Theological Wordbook of the Old Testament*, "Originally this word ['adama] signified the red arable soil. From this it came to denote any cultivated, plantable ground and/or landed property. At times it approaches the meaning of 'home country.'"[17]

Only a small fraction of the world's landmass may be described as arable. Thus, it seems unlikely *'adama* would be used with reference to all Earth's surface or landmasses. More likely, the use of *'adama* in Genesis 6–8 implies that preflood humans limited their range of habitation to the arable (food-growing) land from which they could most easily provide for their nourishment.

The words *'eres* and *'adama* are used 36 and 8 times, respectively, in Genesis 6–8. The context indicates that these two words are used synonymously and interchangeably. It seems they are best understood as referring to the land inhabited and tilled by humans, while land not inhabited by humans would be excluded. As explained in chapter 10, only land inhabited by humans can be damaged by human sin.

In addition to *'eres* and *'adama*, biblical Hebrew includes words for earth and land—*haraba*, *yabbasha*, and *sayon*—that never refer to all Earth's surface or all Earth's landmasses. More than a decade ago, these words were a major subject of a discussion I participated in over the flood's geographical extent. Two theologians who were global flood proponents made the claim that Moses would have used *haraba*, *yabbasha*, or *sayon* for any area smaller than the whole earth. Their argument was that these Hebrew words are never used for the entirety of Earth's landmasses. So, they claimed, *if* Genesis 6–8 indeed refers to only a fraction of Earth's landmasses at least one of the words *haraba*, *yabbasha*, or *sayon* would appear in this passage. For them, the absence of these words served as textual proof that Noah's flood must have been global.

Ironically, *haraba* does appear within Genesis 6–8. It's used in Genesis 7:22 to distinguish the "dry land" from the wet or damp ground.[18] But, even apart from this point, the theologians' argument lacks credibility on both linguistic and contextual grounds.

A quick look at an exhaustive biblical concordance reveals that *'eres* is a familiar noun—the fourth most frequently used noun in the Old Testament, appearing 2,526 times. Only the nouns *yhwh* (LORD), *ben* (son), and *'elohim* (God) appear more often. It may be used for Earth in the cosmic sense, but most often it indicates a specific territory or land area. By contrast, *haraba*, *yabbasha*, and *sayon* are used only rarely in the Old Testament, appearing just 8, 3, and 2 times, respectively. Given Moses's desire to communicate clearly to a broad audience, his preference for familiar nouns over obscure ones makes sense.

Permanence of Oceans' Boundaries

When God says, "Let there be" (*haya*, in Hebrew), what he brings into existence will, presumably (based on his revealed character) remain in existence until his purpose for making it has been fulfilled. This truth is evident in the case of light, Earth's water cycle, the Sun, Moon, and stars, and all the rest of God's creation. The same could be said of Earth's landmass.

The permanence of dry land and oceans as separate features of Earth is affirmed in multiple passages of Scripture, including Job 38, Psalms 33, 65, and 104, Proverbs 8, and Jeremiah 5. All these passages elaborate on creation day three (Genesis 1:9–10). In God's challenge to Job, recorded in Job 38:8–11, he says,

> "Who shut up the sea behind doors when it burst forth from the womb, when I made the clouds its garment and wrapped it in thick darkness, when I fixed limits for it and set its doors and bars in place, when I said, 'This far you may come and no farther; here is where your proud waves halt'?"

The metaphors paint a picture of endurance, implying that once formed, Earth's landmasses would never again be completely submerged as they were in the beginning, *before* God brought forth the dry land. The "thick darkness" seems an obvious reference to Genesis 1:2, which says of the primordial earth, "darkness was over the surface of the deep."

Centuries later, the prophet Jeremiah wrote,

> "Should you not fear me?" declares the Lord. "Should you not tremble in my presence? I made the sand a boundary for the sea, an everlasting barrier it cannot cross. The waves may roll, but they cannot prevail; they may roar, but they cannot cross it." (Jeremiah 5:22)

In this passage, Jeremiah points back to the Genesis creation account, to God's transformation of Earth from a "water world" to a world with both oceans and landmasses (Genesis 1:9–10). He asserts that land would remain an enduring barrier, preventing Earth from ever again becoming a water world.

Psalm 104, the most extensive and detailed of the creation psalms, highlights multiple events of the Genesis 1 creation story, adding some scientific detail. Verses 5–9 speak of God's work on creation day three:

> He set the earth on its foundations; so that it can never be moved. You covered it with the watery depths as with a garment; the waters stood above the mountains. But at your rebuke the waters fled, at the sound of your thunder they took to flight; they flowed over the mountains, they went down into the valleys, to the place you assigned for them. You set a boundary they cannot cross; never again will they cover the earth.

The context of these verses is unquestionably that of creation, not of a later time in history. We see this message repeated in Proverbs 8:29, which says, "He [God] gave the sea its boundary. . . ." This verse, too, affirms that Earth would never again be covered with water.

Global flood proponents respond to this challenge by claiming that Psalm 104 is not a creation account and that Psalm 104:5–9 is all about Noah's flood.[19] Though they are correct in arguing that Psalm 104 is not a chronological narrative that parallels Genesis 1, it is one of the six psalms[20] Old Testament scholars have long grouped together as celebrations of creation, and with good reason. Psalm 104 is the longest and most detailed of these creation psalms. It comments on the events of all six of the Genesis creation days. Compared to Genesis 1, it provides a longer list of God's creation works: the heavens, the clouds, the wind, Earth's foundations, the waters, the mountains, the valleys, the springs, the ravines, the beasts, donkeys, birds, grass, cattle, plants, trees (including cedars and pines), goats, coneys, lions, humans, forests, volcanoes, the sea, leviathan, the Moon, and the Sun. Psalm 104 also describes God's good provision for his creatures. To force-fit a description of Noah's flood into the text of Psalm 104 is to strain its message beyond reason. Verse 24 leaves no room for doubt that Psalm 104 is about God's creation activity:

> How many are your works, Lord!
> In wisdom you made them all;
> the earth is full of your creatures.

Job 38:8–11 and Jeremiah 5:22 pose equally serious challenges to global flood arguments. As with Psalm 104:5–9, to claim that Job 38:8–11 and Jeremiah 5:22 say nothing about creation but, rather, describe Noah's flood lacks exegetical credibility.

New Testament Clues

The apostle Peter twice commented on Noah's flood in his second epistle. 2 Peter 2:5 says, "[God] did not spare the ancient world, but preserved Noah a preacher of righteousness, with seven others, when He brought a flood upon the world of the ungodly" (NASB). In the original Greek text, the two relevant phrases are *archaiou kosmou* and *kosmo asebon*. They literally mean "world of an ancient time"[21] and "world of those destitute of reverential awe towards God."[22] According to Peter, God's judgment came against a world defined by ungodly people, not by the extent of Earth's surface area.

In 2 Peter 3:6, Peter adds, "By these waters also the world of that time was deluged and destroyed." *Tote*, the Greek term translated "of that time," serves as an important qualifier. By modifying *kosmos* with *tote*, Peter communicates to his readers that the world of Noah is other than their world, the Roman world.

Some Bible interpreters who view Noah's flood as globally extensive argue that because the coming "judgment by fire" (referred to in 2 Peter 3:7) will be a global event, the judgment by water (mentioned in 2 Peter 3:6) must be global, also.[23] This argument's flaw is that Peter, here, is speaking not about geography but, rather, about people. Peter points out that this coming judgment from God will fall upon all ungodly people. Today, there are people living in Greenland and Antarctica. At the time of Noah's flood, however, no humans were living in Antarctica or Greenland.

Each time Peter uses the Greek word *kosmos* ("world") with respect to Noah's flood, he qualifies it, and each time with different restrictive adjectives. These qualifiers indicate that *kosmos* is intended to refer to an area less than Earth's total land area.

Peter's understanding of the flood and its purpose no doubt came from conversations with Jesus himself. Matthew 24:38–39 and Luke 17:26–27 record some of Jesus's teaching about Noah's flood and humanity's condition at the time of Noah. In both Gospel accounts, Jesus is speaking about people who have ignored God, focusing entirely on themselves, utterly oblivious to God's call to worship him and serve others. He depicts their spiritual decay as having crept up slowly, but their end comes suddenly, as when Noah's flood "took them all away" (Matthew 24:39) and "destroyed them all" (Luke 17:27). Jesus's reference to "all" refers to the people of Noah's day, apart from Noah and his family.

Jesus also referred to God's similar action toward the people of Sodom at the time of Abraham. As recorded in Luke 17, Jesus said, "The day that Lot left Sodom, it rained fire and brimstone from heaven and destroyed them all."[24]

Once again, he clarifies that God's wrath came against reprobate people and was limited to their location. Lot and his daughters, only a short distance beyond Sodom's city gates, were spared when judgment fell upon the people of that place.

Hebrews 4:3–4 reminds us that God's "works of creation have been finished since the creation of the world." Quoting from Genesis, the writer adds, "On the seventh day God rested from all his works," a clear reference to his *creation* works. To claim that the series of major globally formative events at the time of Noah's flood were responsible for virtually all Earth's geological features and geological history would seem a clear contradiction of this Bible passage.

Limitations on Human Population

Global flood proponents argue that given the long potential lifespans of the preflood people, the human population would have been too vast for one region. So, people must have been globally dispersed. However, they infer that the determining factor in population size at that time was the maximum human lifespan. The biblical text strongly points to another factor in determining the average lifespan of preflood people.

Genesis 4 relates how Cain's murder of his brother Abel led to bolder murders in his family line. Genesis 6 declares that the propensity for violence among Noah's contemporaries was widespread. Perhaps one of the special features of the Genesis 5 genealogy is that it lists those rare patriarchs who died of natural causes rather than from being killed by fellow humans.

The murder rate at the time of Noah may have been so high that, even though humans could live for 900+ years,[25] the *average* lifespan was as short as or even shorter than it is today. Apart from death by murder and/or by other likely causes, average lifespans of 800 years would have resulted in triple the world's current human population even before the end of Adam's life.[26] Not a trace of archaeological evidence suggests that the preflood population grew anywhere near that large. An exceptionally high murder rate could have eliminated the preflood population's need to disperse. It also would have hindered the advance of civilization to a major degree.

Failure to Disperse

God's earliest command to Adam and Eve was that they "multiply and fill the earth" (Genesis 1:28). His plan for humanity called for global occupation so that people could manage Earth's resources for the benefit of all life on Earth, their own and others'. God's later words and actions, as recorded in Genesis

9–11, imply that humans had chosen to carry out their own plan, in defiance of God's.

In Genesis 9:1 and 9:7, God repeats the command first spoken to Adam: "Multiply and fill the earth." In this same passage (verses 5–6), he speaks directly and sharply to Noah about the need to restrain murder. Though the original Hebrew does not explicitly identify who will do the restraining, God tells Noah that he will demand an accounting for the murder of another human. God's intent is that murder will never again become a significant factor in limiting the increase of the human population.

Genesis 11 reveals that several generations after the flood, God's command to multiply was *still* being ignored.[27] Our progenitors remained so stubbornly disobedient in this matter that God finally intervened, forcibly scattering them outward from one region into all Earth's continents and islands for their own survival's sake. God knew that their goal of remaining as one nation in one region with a single capital and a common language constituted a prescription for increasing oppression.

God's intervention may be seen as a form of antitrust legislation. A single government overseeing the whole human population would be tempted toward the kinds of totalitarian governance—and evil oppression—seen in parts of the world today, making life miserable and constantly dangerous for all but those "at the top" of power and wealth. When nations must compete with one another for citizens, competition acts as a check on oppression, corruption, and suffering. The lesson of Babel must not be forgotten.

Genesis 9–11 declares that humanity remained firmly entrenched in a single geographic region. Not until many generations after Noah did humans begin to disperse and settle in all Earth's landmasses. Scientists are now able to use multiple independent methods to determine when the dispersal of humanity and the settlement of all Earth's landmasses began (see chapter 11). Noah's flood apparently occurred a number of generations before this date.

Geographical Markers

Biblical clues to the geographical range of human habitation at the time of Noah can be found in the place-names Genesis records. All the places identified in Genesis 1–9 belong to settlements in Mesopotamia, the Persian Gulf Oasis, and the wider Persian Gulf region. From Genesis 10 onward, the text refers to known places located throughout much of the Eastern Hemisphere.

This sudden shift to a broader geographical focus in the chapters after Genesis 10 strongly suggests that until the time of the flood (and even for a while afterward), humans remained within a relatively limited geographical area. Thus, to fulfill God's purpose in dealing with societal reprobation, God would have sent a catastrophic deluge as far and wide as human society had spread, but not beyond. It seems worth noting that a globally extensive flood would have destroyed virtually all the geographical markers mentioned. The generations of humanity living after the flood would have been unable to locate *any* geographical places mentioned in the Bible from the earlier era, such as the Garden of Eden and its rivers or the land of Nod (Genesis 4:16).

If Noah's flood had destroyed all these possible markers, no reason would remain for Genesis 2:8–14 and 4:16 to be specific and detailed about the location of Eden and of Nod, to the east of Eden. These texts are written as if readers would know and understand such details. The texts convey an "as-you-know" tone in specifying geographical place-names and features.

Floodwaters' Source

Another way Genesis 6–8 sets limits on the extent of the flood is by indicating where the floodwater came from (Genesis 7) and where it returned (Genesis 8). Genesis 7:11–12 says the floodwaters rose up from "the springs of the great deep" (*ma'yenoth tehom rabah*) and flowed down from "the floodgates of the heavens" (*'aruboth hashamayim*). The first of these Hebrew phrases refers to subterranean reservoirs, or aquifers, and the second, to clouds heavy with rain. The quantity of water by these earthly sources measures far short of the quantity required for global inundation—even if the highest preflood hills or mountains were no more than 500 feet above current sea levels.

As with several other desert regions, the Persian Gulf Oasis and Mesopotamia possess geologic features that favor formation of large underground aquifers. Sudden geologic events can cause that water to rise to the surface. And, while rain falls rarely in Mesopotamia and the Persian Gulf, a sufficient gathering of cyclonic clouds from the Indian Ocean, indeed, can bring about an enduring torrential downpour.

Only some of Earth's terrestrial regions hold voluminous underground aquifers. It's geologically impossible for such aquifers to reside under the entirety of Earth's landmasses. Likewise, while 40 days and 40 nights of torrential rain can possibly fall on a few of Earth's terrestrial regions, it cannot fall simultaneously on all.

Some global flood proponents suggest that God delivered most of the floodwaters from outer space. The only known source would be comets. However, the delivery of that much water from cometary bombardment would have destroyed both the ark and its inhabitants. Just as problematic is the fact that the Bible gives no indication of such a source while it does mention rainfall and the springs of the great deep.

In his account of the receding floodwaters, Moses employed four Hebrew verbs: *shakak*, *shub*, *kaser*, and *qalal*, defined as "subsided or abated;" "returned to its original place or condition;" "diminished or lessened;" and "lowered or flowed away," respectively.[28] Their usage indicates that the floodwaters returned to the places from which they came—the aquifers and the atmosphere, where they remain, presumably to this day. If Noah's flood could have been caused by the movement of water from its original location on Earth (and in its atmosphere) to another and back again, it could not have been globally extensive.

Genesis 7 says that torrential rain and the springs of the great deep provided sufficient water for Noah's flood. Apparently, no other significant source of water was necessary. However, Genesis 8 declares that Noah's flood took 335 days to completely recede. For the flood to take so long to completely recede, an additional water source not clearly visible to Noah must have been available. The most likely source would be meltwater from ice and snow on distant mountains. It would have replaced the water from rain and aquifers, which would have flowed out into the oceans within just a few months.

Wind to Assist in the Water's Recession

Genesis 8:1 also tells us *how* God removed the floodwaters from the land: He sent a wind. This drying technique perfectly provides what flooded plains such as those in Mesopotamia and the Persian Gulf regions would require. Water of nearly any depth in such a flat region would flow only slowly toward the ocean. However, a strong wind blowing in the right direction would significantly accelerate the water's movement. Wind also speeds natural evaporation. Thus, wind proves an effective means for removing water from expansive, low-lying plains. A strong, enduring wind would allow the ground to become completely dry (Genesis 8:13–14), free of quicksand and bogs.

With or without strong winds, a quantity of water sufficient to cover the entire globe would have no place to go—certainly not in a period as brief as eleven months. A flood covering one, two, or even three contiguous geographic regions—an area large enough for a sizable human population—definitely could have returned to its place of origin.

The Ark's Landing Place

The claim that Noah's ark came to rest on Mount Ararat has been made by so

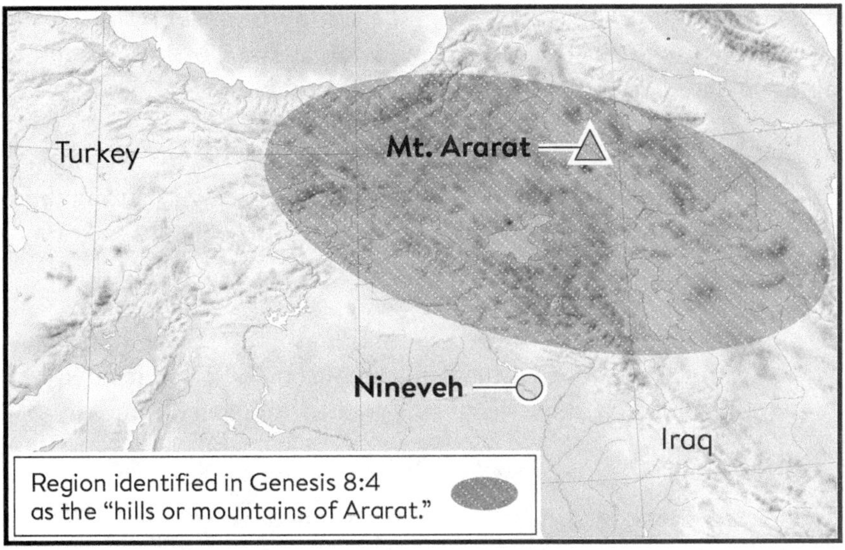

Figure 12.1: Mountains of Ararat
The most likely landing location for Noah's ark would be toward the bottom right of the shaded area. The small solid circle marks the place where the ruins of the ancient city of Nineveh (one of the cities built by Nimrod) were found (Genesis 10:11). *Credit for underlying map: Sémhur, Wikipedia Commons, Free Art License*

many people and so often that it has become "common knowledge." However, this landing spot is not the one identified in the biblical text. Given Ararat's elevation, 5,137 meters (16,854 feet) above sea level, it's no wonder people envision the flood as covering the highest mountains on the earth. However, according to Genesis 8:4, Noah's ark came to rest on the "mountains" (the plural form of *har*) of Ararat. This distinction makes a significant interpretive difference. The entire Ararat range—actually a vast complex of mountain ranges—extends from the north and east of Mount Ararat all the way south to the foothills skirting the Mesopotamian plain. This range encompasses an area measuring more than 250,000 square kilometers (100,000 square miles), with elevations ranging from hundreds to thousands of meters (or feet). The ancient nation of Armenia is geographically defined by the mountains of Ararat.

From Genesis 8:4, we learn that the ark came to rest somewhere within this relatively vast region. Figure 12.1 shows the approximate expanse of the mountains and foothills of Ararat. To search for remnants of the ark's wood on Mount Ararat seems a futile endeavor. The size of the area in which it landed represents one major challenge. Another would be the potential usefulness of its lumber to later residents of the region (see chapter 14).

Noah, Preacher to His World

The book of Hebrews comments on Noah's spiritual ministry during the years he devoted to the ark's construction. The writer of Hebrews tells us Noah preached to the world of ungodly people:

> By faith Noah, when warned about things not yet seen, in holy fear built an ark to save his family. By his faith he condemned the world and became heir of the righteousness that comes by faith.[29]

Peter, likewise, refers to Noah as "a preacher of righteousness."[30]

How would Noah's message have reached all humanity if, in fact, people inhabited many or most of Earth's continents and islands? Noah certainly could have preached righteousness to all Earth's inhabitants if all were living in just one or a few contiguous regions. Again, Noah's flood need not have been global to fulfill God's purpose.

Reasons Not to Emigrate

When God brought judgment upon Sodom and Gomorrah, he told Lot and his family to move out from the region of judgment (Genesis 19:12–23). Global flood proponents argue that God would have done the same for Noah and his family if indeed Noah's flood had not been globally extensive.

Lot's move was a relatively short one. Furthermore, it was not to an uninhabited region. Noah's emigration would have been an extremely distant one through regions inhabited by very wicked people. Risks to himself, his family, his animals, and his possessions would have been enormous. Losses would have been inevitable and of such a proportion as to hamper the relaunch of civilization after the flood.

The restart of civilization in a region never before inhabited would have been much more difficult and certainly more time-consuming. The ark's landing in a region known to Noah and his family with known resources clearly aided and sped up the postflood efforts to reestablish civilization.

What's also worth considering is that Lot had completed his preaching mission (Genesis 19:1–9; 2 Peter 2:6–8) toward a small segment of the human population. For Noah's preaching to reach all people living at his time, he would have needed to remain with the people for the several decades it took him to construct the ark (see chapter 14).

The Crucial Issue of Unchanging Physical Laws

All models and interpretations of Noah's flood as a globally extensive event invoke, by necessity, radical changes in the laws and constants of physics. For example, to generate the degree of plate tectonic movement required to account for the formation of nearly all Earth's geological features within a period as brief as 150 days would require highly accelerated radiometric decay rates—by a factor of roughly several hundred million times.

The authors of a two-volume study, *Radioisotopes and the Age of the Earth* (RATE), rightly acknowledged this problem. In their words, "A change in the decay constant of the order of 10^9 may be required, if the accelerated decay is restricted to the one-year period of the Genesis Flood. . . . A one-year episode of accelerated decay at the time of the Flood may not be enough. Other episodes during Creation week or during the Fall may also be necessary."[31] Even apart from what science would say about the possibility of such an accelerated decay rate, we have words from Scripture to deny it.

Explicit statements in Jeremiah 33:25 and Romans 8:19–22, for example, confirm that the laws and constants of physics have remained unchanged throughout the universe's and Earth's history.[32] Comprehensive measurements and observations by astronomers and nuclear physicists also rule out any such changes. (These constraints are further addressed and documented in chapter 18.)

Consistent Interpretation

Noah's flood was, without doubt, a cataclysmic event. It was the most catastrophic—and potent—warning of sin's consequences ever delivered to humankind. It reduced all humanity to one family from which to repopulate the earth and build a global civilization.

Perhaps because people today have become accustomed to picturing Earth, its peoples, and geographic features in global terms, we tend to forget that humans living during the centuries when the 66 books of the Bible were written had no such images or concepts in mind. The world meant people, the land meant the ground under their feet, stretching from one horizon they could see to the opposite horizon. The highest hills or mountains would be those within view or walking distance (for people and animals) of their homes, farms, and cities.

Throughout the Bible, both Old and New Testaments, all references to and depictions of Noah's flood support this ancient perspective. Not one passage indicates that the story should be understood otherwise. Interpreting all the Bible's passages that are relevant to Noah's flood both literally and consistently leads reasonably to the conclusion that Noah's flood was not globally extensive. At the same time, the passages affirm that Noah's flood destroyed the entire world of ungodly people, sparing only Noah and his family, along with their animals.

In other words, a case can be made from Scripture for interpreting the event as worldwide with respect to humanity, but less than global with respect to Earth's geography. An overview of scientific support upholding the Genesis flood story as a true and trustworthy historical account is presented in the next chapter.

Chapter 13

Scientific Clues to the Flood's Extent

Psalm 19 assures us that God has revealed himself to humanity in two completely trustworthy ways: through the words of Scripture and through his works in nature. The historic Christian doctrine of two revelatory "books" from God is articulated in article 2 of the Belgic Confession.[1] The church has proclaimed for centuries that each of these communications reliably reveals truth, timeless truth, about the Creator and the creation, and these truths are consistent. Stories presented as narratives stand up to historical scrutiny. Stories presented in poetic or artful form are true, as well, and literary devices are clearly identified, as in the case of parables and prophetic visions.[2] The established facts of history and science can be useful in yielding additional details. Such is the case with our study of Genesis 6–9, the account of Noah and the flood of his day.

Misplaced Ridicule

The credibility of the Genesis account of Noah's flood has been challenged or rejected by many people over the past few centuries, regrettably, for all the wrong reasons. Misrepresentation of what the Bible says about this devastating event has been so broadly and loudly disseminated that the true story of Noah and the flood has been obscured.

University of Chicago geneticist Jerry Coyne, a self-identified atheist, describes the story of Noah and the ark as a tall tale that "defies not only geology but reason (how could all of Earth's species, including dinosaurs, stay alive for a year on the ark with a single window?)."[3] Does the story say what Coyne claims it does? University of Mississippi geologist Gregg Davidson, a Christian sympathetic to Coyne's and others' skepticism sees the standard teaching on the flood as the problem. In his words, "All data from nature must be force-fit into the accepted-truth [global] model, no matter how convoluted the resulting story may become."[4] Davidson adds that the popular interpretation calls into question "fundamental laws of physics and chemistry" and, worse yet, means "nature cannot be trusted to tell its own story."[5]

The previous chapter reviews what the biblical text does and doesn't say about the impact and extent of the Genesis flood. This chapter reviews scientific evidence as it relates to the biblical account. The degree of alignment between this evidence and the biblical record provides both corroboration and additional details.

Traceable Flood Deposit?

Many who consider Noah's flood a historical event acknowledge that it must have been large, though less than global. To float a vessel as big as the one described in the text would require a sizable body of water. This is especially the case if Noah can see nothing but water from one horizon to the other from the ark's highest deck. A rough estimate of the quantity would be about six billion acre-feet of water, at a minimum.

Most people presume that such an extensive flood would have left behind sufficient sedimentary evidence for geologists to trace. However, this assumption overlooks the fact of the flood's brevity and ancient date. According to Genesis 7:24, the flood "surged" and remained high for a mere 150 days. Just one year and ten days after it began, it had completely receded.[6] Even the most massive flood of such brief duration would have left no deposit of sufficient substance or endurance to be clearly identified by geologists thousands of years later. One notable exception to this rule would be a tsunami event, hundreds of feet high. Tsunamis of such magnitude leave shoreline deposits that are discoverable even millions of years after they strike. No such evidence exists in the Near or Middle East dating back to less than 100,000 years ago.

Consider the floods that covered the Mississippi and Missouri Valleys in 1927 and 1993. Regions larger than 200 kilometers (125 miles) long by 100 kilometers (60 miles) wide remained under 10–15 meters (30–50 feet) of water for several months. Today, less than 100 years later, the geological and geographical evidence of these floods has already vanished.

As explained in chapter 11, the best estimate of the flood's timing—based on relevant direct and indirect dating methods where systematic errors are known—is that it occurred more than 45,000 years ago. Even if the flood occurred only 20,000 years ago, no geologist would expect to find sufficient evidence for positive identification of a 375-day flood that covered much of the Near and Middle East and parts of eastern Africa by 100+ meters (several hundred feet). It would be just as difficult, if not impossible, to find deposits of such a flood below present-day sea levels.[7]

Given that Noah's flood dates back more than 45,000 years, we can be sure it occurred during the most recent ice age (120,000–12,000 years ago). From the beginning of this ice age to about 45,000 years ago, sea levels fell, at times, to levels far below their current levels—as much as 45–120 meters (148–390 feet) below. These low sea levels led to a radically altered geography for the entire region where the flood most likely occurred. The low levels also imply that deposits from Noah's flood, if any, would now be well below sea level, subjected to episodic rising and falling—effectively erasing them.

The last ice age was characterized by dozens of recurring temperature excursions of major proportions. Deep ice cores from Antarctica, Greenland, and the Alps show that the global mean temperature rose and fell by 8–10°C (14–18°F) on timescales of just a few centuries (see figure 11.4). Such dramatic temperature variability means rapid and widespread reduction and expansion of vast ice sheets, each sheet hundreds of meters (hundreds to thousands of feet) thick. These intermittent expansion-reduction processes would have transformed the geologic features of the land. At least ten dramatic temperature excursions are known to have occurred between the middle and end of the last ice age, even up to 11,000 years ago (see figure 11.4). Thus, erasure of geological evidence for Noah's flood is virtually guaranteed.

Archaeologists and geologists have found several large alluvial flood deposits in the Mesopotamian plain,[8] ranging in thickness from 0.4–2.8 meters (1–8 feet) and all much less than 200,000 square kilometers (80,000 square miles). Because the Mesopotamian plain drains poorly (sloping by less than one foot per mile in the direction of the Persian Gulf), flooding of the Euphrates and Tigris Rivers occurred frequently before construction of dikes and canals in that area. The flood deposits archaeologists and geologists have discovered there are unlikely to have been caused by a flood as large as Noah's.

Supposed Evidence for a Global Flood

Christian leaders who insist that Noah's flood was globally extensive offer what they refer to as "a flood of evidence" in support of their claim.[9] This "evidence," much of it based on the assumption of rapid catastrophic plate tectonics, includes the following:

1. marine fossils and sediment layers near the tops of the world's highest mountains,
2. marine fossils in most of the rock layers in the Grand Canyon walls,
3. deep canyons on all seven continents interpreted as the result of catastrophic carving from the recession of water from a single flood,
4. fossils showing the rapid burial of plants and animals in various parts of the world,
5. rapid deposition of sedimentary layers in several parts of the world,
6. sedimentary layering as the primary composition of Earth's crust,
7. transport of eroded sediments over long distances,
8. rapid erosion between rock layers,
9. lack of erosion between rock layers of apparently different ages,
10. sequences of strata bent without fracturing, said to have been deposited and folded while still wet and pliable,
11. sudden disappearance of Jurassic and Cretaceous dinosaurs,
12. rapidity of the Cambrian explosion,
13. paucity of sodium chloride and other salts in the oceans,
14. multiple reversals of Earth's magnetic field,
15. past formation and breakup of one or two supercontinents,
16. ability of animals and birds to migrate long distances, and
17. the occurrence of a past ice age.

This list—which omits purely hypothetical proposals such as a canopy of water in Earth's atmosphere, exponentially accelerated decay rates of radioisotopes (an isotope is an element with a specified number of neutrons in each nucleus of the element), and a hyper-accelerated supercontinent cycle—looks impressive and compelling to people with no training in relevant scientific disciplines. On closer examination, however, each of these scientific "evidences" does one of two things: either it supports the view that Noah's flood was *not* globally extensive, or it offers no support at all for a global interpretation. Paragraphs under the following seven subheads address these issues.

Marine Fossils and Sediment Layers on High Mountains and Canyon Tops

While it's true that paleontologists and ordinary mountain climbers have found marine fossils and sedimentary layers on the highest reaches of the world's tallest mountain ranges, geophysicists have a clear and firmly established understanding of why. An especially great abundance can be found throughout the Himalayan mountain range because the Himalayas of today are the

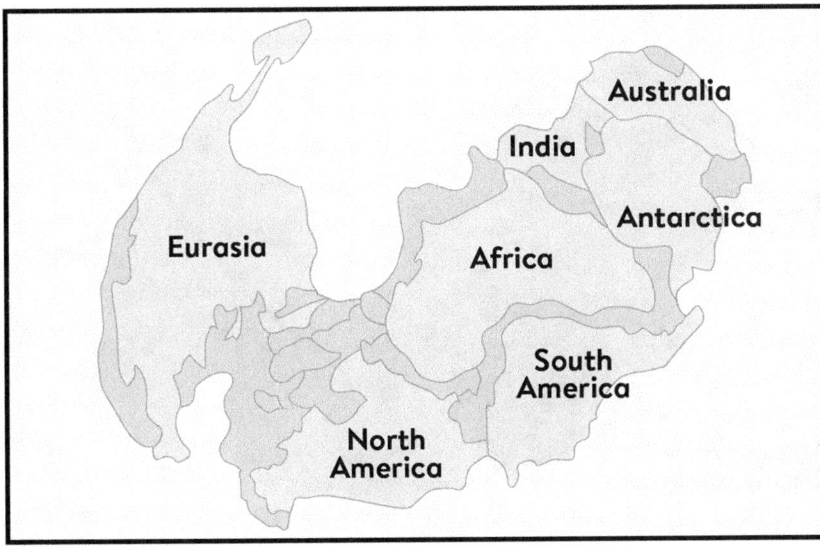

Figure 13.1: Pangea Supercontinent
Outlines of the present-day continents are shown as they were 250 million years ago.
Credit: Kieff

crumpled-up geologic layers that long ago comprised the seafloor between the Indian subcontinent and Eurasia.

Research confirms that the Indian subcontinental crust is geologically distinct from the Eurasian crust. As it turns out, Eurasia and the Indian subcontinent reside on different tectonic plates. Paleomagnetic and seismic measurements show that about 250 million years ago, Earth's continental tectonic plates comprised a single supercontinent, Pangea (see figure 13.1),[10] which began to break apart about 200 million years ago.

Until 140 million years ago, Africa, Antarctica, Australia, and the Indian subcontinent comprised a single continent. A few million years later, the Indian subcontinent and Madagascar split off from that plate. About 100 million years ago, the Indian subcontinent and Madagascar separated. Starting 75 million years ago, the Indian subcontinent began to move, some would say "race," toward Eurasia. Figure 13.2 shows its pathway. For most of this journey, the Indian subcontinent traveled at the astonishing speed of 20 centimeters

Figure 13.2: Movement of the Indian Subcontinent Toward Eurasia
Image credit: NASA; diagram credit: Hugh Ross

per year,[11] four times faster than any other known tectonic plate movement in Earth's history. The thinness of the rigid portion of the Indian subcontinent's plate likely accounts for its speedy migration.[12]

The collision between the Indian subcontinent and Eurasia began about 50 million years ago. Around 23 million years ago, the sea gap between Eurasia and the Indian subcontinent closed. This closure caused the seafloor between Eurasia and the Indian subcontinent to fold upward, gradually forming the Himalayan mountain range. It also pushed up the Tibetan Plateau north of the Himalayas.

The collision between the Indian subcontinent and Eurasia is still ongoing. In fact, the Indian subcontinent currently moves northward by 5 centimeters per year. This ongoing motion means the Himalayas and the Tibetan Plateau continue to rise to ever higher elevations. The tectonic movement also explains the wealth of marine fossils found throughout the Himalayas, the wealth of dinosaur fossils throughout the Tibetan Plateau, and the abundance of sedimentary layers so common throughout that region.

The tectonic movement that formed the Himalayas has also formed several other mountain ranges on Earth. For example, a tectonic collision event explains the discovery of Cambrian explosion marine fossils in the Canadian Rockies. Likewise, North America's vast alluvial plain arose from events of the far distant past. The prairies were once a huge shallow sea where dinosaurs roamed for tens of millions of years. The history of these creatures is responsible to some degree for the region's enormous limestone and fossil fuel deposits, not to mention its numerous pristine dinosaur fossil beds. Several direct dating methods independently show that the prairies emerged from under water more than about 200 million years ago.

Tectonic collision events separated by tens to hundreds of millions of years—not Noah's flood—explain why the marine fossils found in one mountain range differ from marine fossils found in other mountain ranges. These collision events also explain why the sedimentary layers in one mountain range are often distinctly different from the sedimentary layers in other mountain ranges. Additionally, they explain the dissimilarities in terrestrial fossils found in the pushed-up plateaus behind various tectonic collisions.

A combination of multiple time-separated tectonic and erosion events explains the distinct rock layers and fossil sequences observed in various deep canyons. Geologists Carol Hill, Gregg Davidson, Wayne Ranney, and Tim Helble compiled a book in which they and seven other scientists show how the rock layers and fossils in the Grand Canyon refute the hypothesis that most of them were formed by a single flood.[13]

Rapid Erosion and Deposition?

The claimed "scientific evidence" favoring Noah's flood as a global event that shaped nearly all Earth's geological features hinges on examples of rapid geological processes. Global flood proponents apparently presume that mainstream geologists deny the occurrence of rapid geological events and processes. Further, they presume that few or no geological events or processes are gradual. Firmly established geological findings show both these presumptions to be false.

Geologists agree that many sedimentary layers do, indeed, show rapid deposition. This fact doesn't mean, however, that all Earth's sedimentary layers were rapidly deposited and/or all at the same time. Geologists have observed over the past several centuries the gradual deposition of sediments in the estuaries of all Earth's large rivers. They see evidence of similar gradual deposition where rivers long ago ceased to flow. A marine sediment core from off the coast of South Island, New Zealand, reveals a succession of sedimentary layers gradually laid down throughout the past 3.9 million years in a manner that accurately reflects Southern Hemisphere climate changes driven by measured cyclical variations in Earth's rotation axis tilt and in Earth's orbital eccentricity.[14]

Furthermore, not all Earth's geological layers are sedimentary. Some are igneous and others, metamorphic. As much as 12% of Earth's land surface is comprised of metamorphic rock.[15] This type of rock forms when existing rock is transformed, physically and/or chemically, from exposure to high temperatures and pressures over time. This transformation process is gradual, not rapid.

Gabbro rock layers provide another example of gradual deposition. Gabbro rocks form when magnesium- and iron-rich magma slowly cools. Conglomerate rocks form via the gradual consolidation and lithification of gravel and sand. The rounder the gravel clasts in conglomerate formations, the longer that gravel has been eroding.

Marble, one familiar type of metamorphic rock, forms when limestone is subjected to intense heat and pressure over long periods of time. As I explain in chapter 17, the quantity of limestone and marble in Earth's crust is orders of magnitude greater than what could possibly form in a single year.

Global flood proponents are also mistaken in their claim that geological erosion almost always occurs rapidly, within a year or less. While there is evidence for some rapid erosion events, notably erosion generated by landslides, volcanoes, and hurricanes, these events date to different eras and occur at differing rates.

Geological evidence for gradual erosion far supersedes that for rapid erosion events. As with the rapid events, the gradual events exhibit varying rates and occur at different times.

Rapid Tectonics?

As rapidly as the Indian subcontinent crashed into Eurasia, the speed of the collision was not so great as to destroy either the sedimentary layers or the fossils caught in the collision. The same can't be said for the speed of tectonic plate movements that global flood advocates are forced, by their models, to propose. For example, a century ago, they hypothesized that during the relatively brief period of Noah's flood, Pangea split up into the seven continents known today.[16] More recently, they've claimed that the older supercontinent, Rodinia, broke apart shortly after the beginning of Noah's flood, then came back together, as Pangea, during the flood, and then broke apart again late in the flood.[17]

One of many problems with such flood scenarios is that geological evidence firmly establishes the occurrence of at least five supercontinent cycles during Earth's history, cycles of landmasses' coming together and moving apart. Geophysicists have identified and dated the following supercontinents: Kenorland (2.7 billion years ago), Columbia/Nuna (1.9–1.8 billion years ago), Rodinia (1.0 billion years ago), Pannotia/Vendia (0.60–0.55 billion years ago), and Pangea/Gondwana (0.25 billion years ago).[18]

The known cycle compels global flood proponents to force-fit not just one or two supercontinent formations and breakups into the 375-day duration of the flood, but at least five. They must squeeze 40,000 kilometers (25,000 miles) of lateral tectonic plate movement into less than a single year!

Global flood proponents must invoke not only extreme lateral tectonic movements but also extreme tectonic uplift. Earth's present ocean basins are too deep and its mountain ranges too tall for water to cover the whole of Earth's surface. Global flood proponents respond by proposing a sudden uplift of all Earth's ocean basins at the beginning of Noah's flood. They say this movement would have caused tsunamis hundreds of meters high to rush over all Earth's landmasses, which they say were significantly lower in elevation at that time.[19] They also claim that later in the course of Noah's flood sudden and dramatic tectonic uplifts greatly increased the elevations of continental mountain ranges and plateaus, while equally sudden and dramatic tectonic downfalls deepened ocean trenches and basins.[20] They calculate that within just a few months, Earth's mountains must have been pushed up by thousands of meters (feet) and ocean floors pressed down by still more thousands of meters (feet).

According to global flood models, all this vertical and lateral tectonic activity during Noah's flood was accompanied by equally extreme erosion forces. Thus, virtually all Earth's geological formations, they say, could be attributed to the 375-day duration of Noah's flood. The same models also predict that virtually all Earth's deposits of coal, oil, natural gas, gypsum, and limestone plus nearly all Earth's fossils formed during this same period (see chapter 17).[21]

According to Genesis 8:2–4, only 150 days after the floodwaters rose, Noah's ark came to rest on the hills of Ararat. Any major tectonic activity after that time would have destroyed the ark and killed its occupants. The same would have been the case during the earliest stage of the deluge. The energy release from 40,000 kilometers of plate tectonic movement within 150 days or less not only would have killed everyone on board Noah's ark, it would have destroyed all fossils, including marine fossils on the Himalayan mountains, evaporated all Earth's surface water, and turned Earth's surface into a molten sea.[22]

The speed of this tectonic movement would average at least 185 meters

(607 feet) per minute. This rate of movement can be put into perspective by contrast with that observed in the 2011 earthquake that struck along the northeastern coast of Japan. During the 5- to 6-minute duration of that magnitude 9 earthquake, a 500-mile strip of Honshu Island, from Tokyo to Sendai, moved laterally by as much as 8 feet.[23] The *average* rate of continental movement demanded by global flood models is at least 400 times greater!

Global flood models imply that within less than 150 days, Earth would have been battered by the force of about 700,000 consecutive magnitude 11 earthquakes. (The energy release of a magnitude 11 earthquake is 100 times greater than that of a magnitude 9 earthquake.) However, the situation would have been far worse.

The structure of Earth's crust is not conducive to continual steady movement of crustal plates. Instead, the plates lock, then lurch, lock again, and lurch again. Thus, earthquakes are episodic. This means global flood models propose what amounts to thousands of magnitude 13 earthquakes (or tens of thousands of magnitude 12 earthquakes), accompanied by countless volcanic eruptions and volcanic island collapses.

The effects of such concentrated volcanism, erosion, and tectonics, had they occurred a few thousand or a few tens of thousands of years ago, surely would be observable and measurable by geophysicists today.[24] Fault lines scarring the planet would be vastly larger and more numerous than what we see. However, owing to the extreme energy release, we wouldn't be here to see them.

Perhaps the greatest of many challenges to this scenario's plausibility is the amount of energy required to make these radical alterations to the planet in such a brief geologic moment. The energy requirement exceeds by millions of times what the laws of physics allow. So, global flood proponents assert that certain constants of physics were altered during the flood by factors ranging from millions to billions of times.

Specifically, these advocates propose that normally long-lived radioisotopes, such as potassium-40, thorium-232, uranium-235, and uranium-238, must have decayed millions, if not billions, of times faster during Noah's flood than before and after it.[25] This ultrarapid decay, they say, would have generated sufficient energy to compress all the geological transformations caused by the flood into less than a year.

Rapid Radioisotope Decay Problems

As noted in the previous chapter and explained at length in chapter 18, the Bible teaches that throughout the history of the universe and Earth, the physical laws and constants God put into effect "in the beginning" have not changed. Thus, from a biblical perspective alone, vastly accelerated radiometric decay rates during Noah's flood would be ruled out. From a scientific perspective, multiple independent lines of evidence refute the notion that radiometric decay rates were radically different during the flood.

Astrophysical evidence: For astronomers, measuring the radioisotope decay rate is direct, straightforward, and unambiguous. Astronomers routinely observe the spectra of the Sun, stars, nebulae, galaxies, and quasars at distances ranging from zero, to tens, hundreds, thousands, millions, and billions of light-years. Given that light's velocity has not changed, these observations reveal radioisotope abundance levels plus the abundance levels of the decay product isotopes (daughter nuclides) throughout cosmic history. Simple calculations transform these abundance levels into radioisotope decay rates.

Hyperfine split spectral lines within the same spectra demonstrate that the velocity of light has, in fact, remained constant throughout cosmic history.[26] Einstein's special relativity provides a second independent verification that light's velocity has not changed.[27] $E = mc^2$ establishes that even a tiny change in light's velocity would have dramatically altered the luminosity of stars—something astronomers would have observed. Furthermore, the change in the Sun's heat and light flow would have wiped out all Earth's life. Therefore, the spectra of the Sun, stars, nebulae, galaxies, and quasars establish the constancy of radioisotope decay rates throughout the history of the universe and Earth.

Some global flood proponents have speculated that Earth's physics may be radically different from the physics of the rest of the universe. However, if that were the case, observable effects would affirm such a radical difference. More importantly, the existence of physical life would be impossible. Henry Morris himself, one of the leading global flood proponents of the twentieth century, explained the impossibility of this difference in one of his books.

Ice and sediment core evidence: Another direct confirmation of the constancy of radioisotope decay rates comes from analysis of the annually deposited layers seen in deep ice core and sediment core drills. (Yet another confirmation comes from the annual growth layers in tree trunks and coral reefs.) Given a sufficiently global sample of ice and sediment cores, the layers in those cores can provide a test of whether or not Noah's flood was globally extensive.

Radioisotopes in three Antarctic ice cores—Dome C, Dome Fuji, and Vostok—provide an annual record of the past 800,000, 720,000, and 420,000 years, respectively.[28] One ice core in Northern Greenland, NGRIP, and two ice cores in central Greenland, GRIP and GISP2, similarly yield an annual record of radioisotopes for the past 123,000, 105,000, and 105,000 years, respectively.[29] These cores leave no doubt that radiometric decay rates have not varied through the past 800,000 years and that Noah's flood fell short of reaching either Antarctica or Greenland.

Concerning the ice core records, global flood proponents have argued that the Antarctic and Greenland ice core layers are not annual layers but, rather, are laid down multiple times per year. They also argue that only one ice age has occurred in Earth's history, an ice age that peaked about 200–500 years after Noah's flood.[30] They further claim that all, or virtually all, the ice covering Antarctica, Greenland, the Himalayas, the Alps, and other parts of Earth were laid down *after*, not before Noah's flood.[31] Global flood proponents hypothesize that many melt events followed by freeze-ups occur every year throughout all of both Antarctica and Greenland. In that case, they maintain, the cores provide evidence not of the past 800,000 and 123,000 years of Earth's climate history, but, rather, only of the past few millennia.

To support such a hypothesis, global flood proponents cite the 1990 and 1992 discoveries of a B-17 bomber and two P-38 fighter planes that had crash-landed in southern Greenland in 1942. The three planes were found under 250 feet of ice.[32] However, glaciologists were not at all surprised by this find, given the well-known and often-observed fact that glaciers in southern Greenland are fast-moving (the planes were found three miles from their original location), unstable, and located at low elevations. For that area, the subsequent burial of aircraft to a depth of 250 feet in 50 years' time appears perfectly reasonable.

All six of the ice cores mentioned previously were drilled in high-elevation, high-altitude, ice-stable regions of central Antarctica and central Greenland, hundreds of kilometers (miles) from the nearest seacoasts. The sites from which these six ice cores were taken are so cold in both summer and winter that melt events do not occur. Meanwhile, cryologists (scientists who study snow and ice) offer at least three additional reasons for certainty that the ice core layers are annual layers. First, the distinct dust signatures of known volcanic eruptions are embedded within the layers. In each ice core, researchers found the unique dust signature of Krakatoa's eruption in 1883 and that of Vesuvius's eruptions in 79, 472, 512, 968, 1037, 1139, 1631, and 1944 AD. A simple counting of layers and comparing the numbers with known eruption dates confirmed that each of these layers corresponds to one year.

Further verification comes via cyclical variations in the eccentricity (a measure of variance from circularity) of Earth's orbit around the Sun, clearly observable in the fluctuation of the ice core layers' thickness (see figure 13.3).

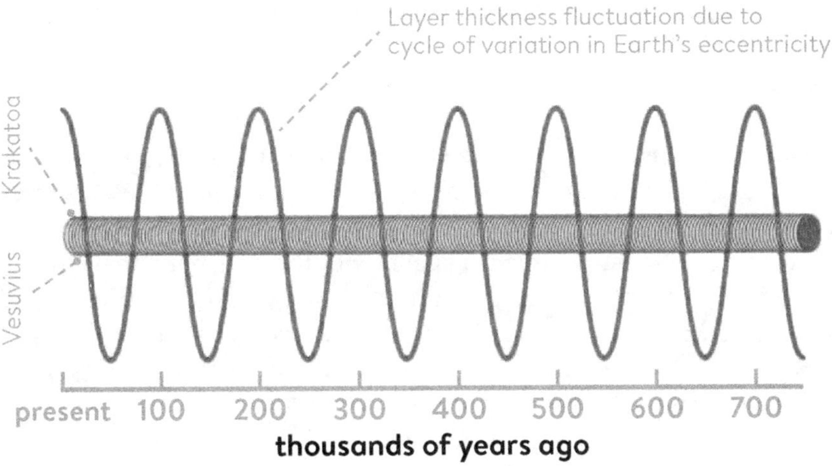

Figure 13.3: Earth's Orbital Eccentricity Cycle Seen in the Dome C Ice Core
The 800,000 layers in the Dome C Antarctic ice core reveal layer thickness variations (schematically drawn) corresponding with the 100,000-year periodic variation in the eccentricity of Earth's orbit. Also evident in the annual layers are dust signatures of volcanic eruptions that occurred throughout recorded history. *Credit: Reasons to Believe*

Seasonal changes affect layer thickness. As Earth's eccentricity increases, the seasons become proportionally warmer. Gravitational interactions between Earth, Jupiter, and Saturn generate an approximately 100,000-year periodic cycle in the eccentricity of Earth's orbit.

The 800,000 ice core layers in Dome C show eight complete cycles of variation in Earth's orbital eccentricity. The Dome Fuji and Vostok ice cores show seven and four complete cycles, respectively. Thus, the law of gravity and Isaac Newton's laws of motion dictate that the Antarctic ice core layers are annually deposited layers. They also definitively establish that a minimum of eight ice ages (episodes of ice maximum) have occurred in Earth's relatively recent history.

As if more confirmation were needed, a marine sediment core from near the coast of South Island, New Zealand, provides evidence of Southern Hemisphere climate changes throughout the past 3.9 million years.[33] The sediment layers observed in this core align with the observed variations in Earth's eccentricity and signature events in the three deep Antarctic ice cores. In this way, we can be sure that the three ice cores really do provide a faithful record of the past 800,000 years of Earth's history and affirm that Noah's flood never covered Antarctica.

Tree ring evidence: A tree's growth rate varies in response to seasonal climate changes, resulting in visible growth rings. Each ring represents one year in the tree's life. Bristlecone pine chronology of the American Southwest provides at least 8,500 years of tree ring data (with the possibility that 3,000 years may be added).[34] European oak and pine chronology, a composite of work done in Germany and Northern Ireland, now extends over the past 11,000 years.[35] A Palmer's Oak in Jurupa Valley, California, which reproduces through generating exact clones near its original trunk, shows via its ring data that it may be the world's oldest known living plant, at 15,600 ± 2,500 years.[36] Here, too, volcanic eruptions in recorded history have left their signatures in specific tree rings, verifying that each ring corresponds to one year.

As with layers in the ice and sediment cores, the annual rings of these ancient trees contain radioisotopes, and analysis of these isotopes confirms that radiometric decay rates have not varied at any time throughout the past 15,000 years. This study also tells us that the American Southwest and Western Europe have experienced no flood of continental proportions within the past 15,000 and 11,000 years, respectively.

Biological evidence: It's important to note that a change in the radiometric decay process would not be good for life. Had the ark's passengers been exposed to vastly more rapid radioisotope decay, they would not have survived. Here's why. All life-forms contain potassium-40. Plants and animals also contain small amounts of thorium-232 and uranium-238. Accelerating the decay rates of these radioisotopes by factors of a million to a billion would have incinerated all life, on and off the ark.

With a view to the deadly impact of greatly accelerated decay rates, some global flood proponents have suggested that a great depth of water underneath the ark would have protected the life on board.[37] While this water might have offered some protection for life from radioisotopes in Earth's crust below the water, it would have offered no protection from radioisotopes in the animals, plants, and food on board the ark. In fact, the water itself would not have survived in liquid form. Radioisotopes in the crust would have released more than enough heat to vaporize all the water on Earth's surface. Such a scenario does not match the message of Genesis 7 and 8.

Russell Humphreys, a physicist who participated in the RATE study (mentioned in the previous chapter), recognized this vaporization problem and proposed a way around it. He hypothesized that "cooling of the earth by the expansion of the cosmos may have occurred simultaneously with the heat produced by accelerated decay."[38] This scenario is ruled out, however, by the fact that cosmic expansion cools only the cosmic background radiation (the radiation left over from the cosmic creation event) and has no measurable impact on Earth's surface temperature. Astronomers measure only a minuscule acceleration of the cosmic expansion rate (less than half a percent) over the past 4 billion years of Earth's history. Throughout the past million years, the cosmic expansion rate measures constant to within 1%.

Geophysical evidence: Earth's mantle and core, which comprise 99% of Earth's volume, are richly endowed with potassium-40, thorium-232, uranium-235, and uranium-238. Accelerating these radioisotopes' decay rates by factors of a million to a billion would have turned the entirety of Earth into a molten blob. Even if decay rates quickly returned to their present values, Earth's iron core would be totally liquid to this day. Seismic measurements show that this is not the case. Earth has a solid iron core that extends 1,221 kilometers (760 miles) outward from Earth's center. Further, if Earth's core were totally liquid at the present time, Earth's magnetic field would be weak, less than a tenth of its currently measured value of 0.45 gauss.

As more problems emerge with the proposed acceleration of the radioisotope decay rate, attempts to escape them stretch credulity beyond the breaking point. One of these attempts invokes a dramatic volumetric expansion of the Earth and Sun during Noah's flood.[39] The claim is that such rapid, major volume increases, caused by a much lower value of the universal gravitational constant, would have cooled the ark and its inhabitants. However, the proposed volume increases would have extinguished the Sun's nuclear furnace and ripped Earth apart.

The Water Problem

Global flood scenarios face yet another scientifically insurmountable challenge: the insufficiency of surface water to cover the whole planet. While there is at least three times as much water—in the form of hydrated minerals—deep in Earth's mantle than in Earth's oceans, lakes, and rivers,[40] the tectonic energy required to extract water from these minerals and deliver this water rapidly to Earth's surface would have so altered Earth's atmosphere as to destroy the ark and kill all its life.

To cover all Earth's surface with water would require more than four times the sum of all water currently present in Earth's oceans, lakes, and rivers, atmosphere, *and* crust. Even if all oceans, seas, and lakes were shallow and mountains and plateaus no more than a few hundred meters (several hundred feet) high, the quantity of available water would still be insufficient.

Many global flood proponents claim that a hypothetical preflood canopy enveloped Earth and provided the water needed to cover all of Earth's surface. They also invoke such a canopy to explain the long lifespans of some preflood humans (see the Genesis 5 genealogy). However, for many reasons, most global flood proponents who hold advanced degrees in science no longer appeal to this hypothesis (see sidebar Canopy Hypothesis Problems).

Marine Salt Problem?

Another claim by global flood proponents is that Earth's oceans would be much saltier if there had been more than one major flood in Earth's history. This claim suffers from the erroneous assumption that water runoff from Earth's landmasses steadily adds salts to Earth's oceans without any of those salts being returned to Earth's landmasses. Plate tectonics pushes up some of Earth's seafloors to make new landmasses. Seawater trapped in these landmasses evaporates to form salt beds. This tectonic activity explains why many salt deposits are found on Earth's landmasses. Salts are continually cycling from the continents and islands to the oceans and back to the continents and islands.

Magnetic Reversals

Global flood proponents allege that Earth's magnetic field has steadily declined since Noah's flood ended.[41] Although they acknowledge that 20 or more magnetic field reversals are known to have occurred in Earth's history, they say that *all* transpired during the 375 days of Noah's flood. They hypothesize that during the flood, these magnetic reversals were generated by rapid "strong flows of the fluid in Earth's core."[42] However, the sudden, violent energy releases needed to produce rapid, strong flows in Earth's core—energy releases of sufficient magnitude to produce 20+ magnetic reversals within a single year—would have destroyed the ark and killed all its inhabitants.

Agricultural Challenge

Genesis 8:11 says that, while the waters of Noah's flood were still receding, the dove Noah released from the ark plucked an olive leaf and brought it back to him. No olive tree or its leaves would have survived the destructive forces of

Canopy Hypothesis Problems

The canopy hypothesis posits that a thick water canopy existed in or above Earth's atmosphere, shielding life from radiation, greatly extending human longevity, and creating a warm, humid environment to augment Earth's biomass and biodiversity. It claims that at the beginning of Noah's flood the canopy collapsed, suddenly inundating Earth's entire surface, destroying all terrestrial life—except life on the ark—and transformed virtually all Earth's topography.

In reality, a water vapor canopy with sufficient water to cover all Earth's landmasses would have either dissipated to interplanetary space or trapped so much of the Sun's heat (water vapor is a powerful greenhouse gas) that no ice, liquid water, or physical life of any kind could have existed on Earth's surface. For these reasons, some global flood proponents hypothesized that a liquid water or frozen water canopy once surrounded Earth. However, gravity would not permit such a sizable, weighty canopy to either form or remain stable.

It's highly debatable whether an increase of surface heat and humidity from this hypothesized canopy would have increased Earth's biomass and biodiversity. In tropical places today, heat and humidity sometimes rise to levels deadly to certain plant and animal species, as well as to humans. Even if biomass and biodiversity had benefited, the enhancement would have been minimal.

Further, while water in the atmosphere can provide some protection from ultraviolet radiation, it can't impede hard cosmic radiation. It's this type of radiation that limits human lifespans today to about 120 years (see Genesis 6:3).

A canopy sufficient to play a major role in covering all Earth's surface with water would block out the visibility of the Sun, Moon, and stars. All canopy speculations create a conflict with Genesis 1:14–16, which says that on creation day four, the Sun, Moon, and stars became visible to creatures on Earth's surface as distinct objects in the sky.

40,000 kilometers of lateral tectonic movement, thousands of meters of vertical tectonic movement and erosion, and violent volcanic eruptions all packed into a single year's time. The fact that olive trees grow at low altitudes also limits the elevation of the waters and of the ark's landing place.

In Genesis 9 we read that Noah began productive agriculture soon after leaving the ark. Such an endeavor would have been impossible if extreme tectonics and erosion had totally transformed Earth's geological features. Aftershocks from 40,000 kilometers of tectonic movement would have made agriculture and even basic survival impossible for many decades after Noah's flood.

Tsunami Challenge

Thousands of kilometers (miles) of tectonic movement squeezed into less than the flood's initial 150 days would have resulted in massively devastating erosion events—landslides and volcanic cone collapses, both above and below the water level. These events, in turn, would produce an especially destructive category of tsunamis. For comparison, the 1958 Lituya Bay (Alaska) landslide caused a tsunami wave along some shorelines with a runup height of 1,700 feet. Thousands of kilometers (miles) of tectonic movement within less than 150 days would generate not just one but thousands of tsunamis larger than the one produced by the Lituya Bay landslide. The ark and its inhabitants could not have survived such an onslaught of tsunamis.

Likely Extent of Noah's Flood

As chapter 11 explains, Noah's flood seems likely to have occurred during the most recent ice age. Figure 11.4 shows that several temperature fluctuations of 8–10°C occurred in Greenland during the period. Analysis of tree rings and foraminifera (single-celled organisms that build shells from seawater minerals) reveal that a similar degree of temperature change occurred farther south, though at a somewhat slower rate. Such dramatic temperature upsurges would have melted large fractions of Earth's ice coverage, ice sheets hundreds of meters (hundreds to thousands of feet) thick. Detailed reconstruction models affirm that rapid melting of vast ice sheets did, indeed, occur several times between 80,000 and 19,000 years ago.[43] This melting would have led to widespread flooding as well as rising of sea levels by tens to nearly a hundred meters.

An enormous ice age melt event would have substantially augmented the waters from the 40-day downpour and upwelling of water from underground aquifers ("springs of the deep"). Geologist Carol Hill explains that "the duration of the rain (forty days and forty nights) could have been caused by the stalling of a Mediterranean cyclonic front over the Mesopotamian area in combination with maritime air masses moving up from the Persian Gulf and Arabian Sea/

Indian Ocean."[44]

Another possibility is that tectonic activity, erosion, and/or dust deposition caused a berm (a raised barrier) to form in the Hormuz Strait, at what is now the mouth of the Persian Gulf.[45] Evidence for the existence of such a berm has been found at a depth of about 85 meters (280 feet).[46] This berm was likely much higher before Noah's flood, but if it were breached by an earthquake or tidal wave, it would have brought a rush of Indian Ocean water into the Near and Middle East.

The much lower sea levels during the previous ice age, by as much as 120 meters (390 feet), would have contributed to the expanse of the flood. Even a 35-meter drop in sea level would turn 53% of the surface area of the present Persian Gulf into dry land and make it vulnerable to flooding.[47] A 120-meter sea level drop would leave all the Persian Gulf, some southern and northern parts of the Red Sea, and portions of the eastern Mediterranean dry and susceptible to flooding.

Taking all these factors into account, Noah's flood could have been far more extensive than traditional "local flood" models propose. Typically, these models imply that Noah's flood covered only the Mesopotamian plain at its current size. A flood of greater expanse would have been necessary to accomplish God's purposes for the flood—to take out all humanity and their *nephesh*, except those within the ark.

How much of Earth's land humans and their animals occupied during the years before Noah's flood can't be precisely determined. A literal, historical interpretation of Genesis 10 and 11 indicates that not until *after* the flood did the world see an aggressive, nearly simultaneous human migration to and through eastern Africa and the Near and Middle East into southern and northern Europe, Asia, Australia, New Guinea, and Tasmania (see chapter 11).

The postflood population expressly chose "to not be scattered over the face of the whole earth" (Genesis 11:4). This statement suggests that they likely remained in roughly the same regions where they lived before the flood. In this case, the maximum extent of Noah's flood would have included all of what is now the Persian Gulf, all of Mesopotamia, much of southern Arabia, the coastal regions of southern Iran and Oman, parts of the Gulf of Oman, the lowlands bordering all sides of the Red Sea, much of the southern and northern portions of the Red Sea, the lowlands of Egypt, Lebanon, Turkey, and the eastern Mediterranean.

If Noah's flood occurred during the last ice age, regions of human settlement likely expanded and shrank in response to the dramatic temperature changes that caused huge ice sheets to melt and reform (see figure 11.4). If a major source of the water for Noah's flood was the rapid melting of massive quantities of ice and snow, it seems reasonable to conclude that Noah's flood occurred during one of the episodes within the last ice age when ice coverage was at a maximum, or nearly so. Given what we can determine about Earth's surface temperature at such times, people had likely moved into the warmer, lowland regions of eastern Africa and the Near and Middle East. If the climate turned exceptionally frigid, people may have migrated all the way into the Persian Gulf region, southern and central Mesopotamia, and lowland regions bordering the Gulf of Oman. This latter area would seem to represent the minimal extent of Noah's flood.

"As Far as the Eye Can See"

During the flood, the ark likely floated somewhere in the region of the Persian Gulf or the Mesopotamian plain. If so, the adjacent high country of Arabia and Persia would have been more than 200 kilometers distant from the ark for several weeks. The current elevation of these mountains is less than 500 meters—too low to be seen from distances of 100+ kilometers (due to the curvature of Earth's surface).

Given the likely position of the ark at the height of the flood, the extent of the waters would match the description given in Genesis 7:19–20. From one horizon to the opposite horizon, all Noah could see from his perspective on the top deck of the ark was water and clouds.

Removing a Barrier to Truth and Faith

The notion of a "geologically impossible" global flood is one major obstacle skeptics will point to as a reason to reject the Bible's accuracy and authority. As we've seen in this chapter, though, God's book of nature (the established scientific record) clearly and firmly contradicts the popular claim that Noah's flood was globally extensive and responsible for virtually all Earth's geological history and features.

This chapter also provides a scientific interpretation of the Genesis account of Noah's flood that is theologically sound and biblically consistent. It shows that scientific evidence supports the plausibility of an inundation so extensive as to take out all humans and their *nephesh*, other than those aboard Noah's ark. In doing so, I hope to have helped remove a major barrier to belief in the truthfulness of Scripture. The next chapter addresses another challenge to the plausibility of this biblical story: the necessity and feasibility of the ark.

Chapter 14

The Necessity and Feasibility of the Ark

Many skeptics have mocked the biblical account of Noah's flood based on its description of the ark. They reject the idea that Noah could have constructed such a large ship, even if he had a large labor force. Further, they argue that even if Noah had somehow managed to build the ark, the idea that it could have been seaworthy for so many months and under turbulent conditions appears laughable.

Global flood proponents make an issue of the ark in their claims, too. They say that if Noah's flood were less than globally extensive, no such vessel would have been necessary. God could have moved Noah, his family, and their animals to safety. He could have simply guided them to a safe location.

These challenges are thought-provoking, but through careful reading of the text and the application of reason, the seeming "impossibilities" or "reasonable alternatives" can be reassessed. The story and its claims can and do become plausible and rationally defensible. God's instructions to Noah (Genesis 6:14–16), for construction of a massive vessel give a strong hint about the magnitude of the impending catastrophe, and God used the building process to amplify Noah's preaching to his contemporaries.

Why an Ark?

Before explaining the specifics of the ark's construction and seaworthiness, I'll address the matter of the vessel's necessity. If ancient humanity had failed to disperse and remained in a relatively limited region of Earth, why wouldn't God have dealt with Noah's situation as he did with Lot's, recorded in Genesis 19? Why would God ask Noah to invest so much time and effort, not to mention such an abundance of resources, when he could have evacuated Noah and his family and animals from the region? Walking would have been much easier than building a ship. Why not simply instruct Noah to pack up and then guide him to a region far away, where he and his family and livestock could settle down and be out of harm's way?

Several reasons come to mind. First, Scripture reveals a clear pattern of prophetic warning in advance of God's judgment. As God raised up Noah and Lot to serve as his prophets of impending judgment, he did the same for Moses, Elijah, Isaiah, Jeremiah, Ezekiel, Daniel, Jonah, and other righteous prophets living in a wicked society. God always sends a messenger to warn of looming judgment. He sends a prophet and gives that prophet a pulpit. He offers deliverance to those who will repent. Thus, he affirms to those being judged and to future generations that divine judgment is both just and essential.

Noah's platform was the scaffolding around his massive construction project. The efforts of a distinguished patriarch, his family, and potential hired workers to build an enormous ship in the middle of a desert plain with scant rainfall no doubt commanded widespread curiosity. It's easy to envision people making special trips to witness for themselves one of the wonders, if not *the* wonder, of the ancient world.

Noah's unwavering dedication to such an immensely challenging project over what may have been as much as a hundred-year period would have heightened the drama. As crowds gathered to gawk, Noah preached (Hebrews 11:7; 2 Peter 2:5), warning onlookers of impending doom and calling them to repent of their evil ways. He probably offered safe passage to anyone willing to heed his warning, and the enormity of the ark would have demonstrated the sincerity of his offer.

The New Testament describes Noah as "a preacher of righteousness" (2 Peter 2:5). As an example of his obedience to God, Noah "waited patiently" (1 Peter 3:20), and his faith "condemned the world" (Hebrews 11). He may have been able to build the ark more rapidly had he devoted less time to preaching. However, the gravity of the coming disaster likely compelled him to offer repeated warnings to any who might listen.

Noah's situation contrasts with Lot's in several significant ways. Noah was faced with pervasive reprobation. Lot encountered the evils of Sodom and four adjoining cities in a small plain. Lot was not a native to Sodom. He moved there as an adult and served for some time as a "judge," according to Genesis 19:1, 9 (ineffectively, it would seem). Lot had a ready platform from which to preach. Sodom and its four adjoining cities were small in both area and population. One short trip would remove Lot and his family from peril and from a region less than essential to humanity's survival.

Limiting the Damage

If God had chosen to move Noah, his family, and his animals, at least some of Noah's contemporaries would likely have chosen to follow. Such a spreading of reprobate people into a larger region would have required a still more devastating flood. By allowing Noah to remain in his place, God minimized the extent of devastation.

The construction of a large ship near the center of human habitation kept Noah's people focused in and on that locale. The people's response demonstrated the depth of societal reprobation. The utter rejection of Noah's preaching and the mockery of his project removed any doubt about the necessity of God's act of judgment, not only for the generations that immediately followed Noah's flood but also for all generations to come. As Jesus commented in one of his sermons, "In the days before the flood, people were eating and drinking, marrying and giving in marriage, up to the day Noah entered the ark; and they knew nothing about what would happen until the flood came and took them all away."[1] It seems the people had completely shut out all interest in or concern for Noah's message and actions.

The obvious foreshadowing provided by Noah's flood as it relates to New Testament teachings on salvation, baptism, and grace may be considered another reason for God's allowing Noah's story to unfold as it did. The rescue of Noah and his family in an ark amid a flood provides an analogy to baptism (1 Peter 3:20–21), symbolizing spiritual rebirth in Christ and baptism in the Holy Spirit (2 Corinthians 1:21–22; Ephesians 1:13–14; 1 Peter 3:21). For Noah and his family, a long walk to safety could be interpreted as an illustration of salvation by works. The rescue of Noah and his family in the ark beautifully illustrates salvation by grace alone. Salvation by grace is seen even in the miracle of Noah's successful construction of the ark. Without God's detailed instructions and daily assistance, Noah could never have accomplished the task.

Some theologians—Christian, Jewish, and Islamic—have argued that the doctrine of salvation by divine grace, alone, is taught only in the New Testament, not in the Old. The story of Noah's flood serves as a counter to that argument. In addition, it illustrates the biblical message that people committed to their own ways, not God's, see the wisdom and actions of God as foolishness. The dismissive response of Noah's contemporaries to his God-ordained building project powerfully underscores this principle.

The Gift of God's Design

This ark was no ordinary boat. In Genesis 6:14–16 we see that God gave Noah specific instructions on the ark's dimensions, construction materials, architecture, and floor plan.

God commanded Noah to construct the ark. He didn't say or imply that he would miraculously intervene to construct the ark. The wording of God's command implies that Noah possessed or had access to the resources, skills, and technology needed to build the ark. It seems highly unlikely, however, that Noah could have constructed a vessel of such seaworthiness and livability without God's specific plans and instructions. It also appears unlikely that Noah could have accrued all the building materials and labor necessary for the ark's construction without God's help.

God instructed Noah to build a vessel 300 cubits in length, 50 cubits in width, and 30 cubits in height. The "cubit" is an ancient Near Eastern unit of measure based roughly on the distance from a grown man's elbow to the tip of his longest finger. According to some Near Eastern sources, a "common cubit" was 44.4 centimeters (1.46 feet),[2] and a "royal cubit" measured 52.92 centimeters (1.73 feet).[3] According to the Talmud and several pre-Christian rabbis, the biblical cubit was either six or seven handbreadths, which added up to 44.87 centimeters (1.472 feet) and 52.55 centimeters (1.724 feet), respectively.[4]

Using the shorter of the biblical cubits, the size of Noah's ark would have been 134.6 x 22.4 x 13.5 meters or 442 x 74 x 44 feet (about the size of a World War II light cruiser). By the longer cubit, Noah's ark would have been 157.7 x 26.3 x 15.8 meters or 517 x 86.2 x 51.7 feet (about the size of a World War II heavy cruiser).

Based on studies of the largest wooden ships constructed in modern times, many scholars argue that for Noah to have built a wooden ship the size of a World War II naval cruiser would be impossible. In response to this apparent dilemma, some interpreters have proposed that at the time of Noah, people were using a numbering system other than the decimal system used throughout the world since the time of Moses.

Christian geologist Carol Hill, for example, has suggested, "It may be that the dimensions of the ark were never converted from a sexagesimal system into a decimal-based numbering system. If one arbitrarily divides the ark dimensions by either ten or six, one comes up with a size more compatible with boats known to have existed in Jemdet Nasr time [the time of the ancient Akkadians/Sumerians]."[5] This suggestion seems unrealistic, however, given that it would translate into an ark too small to accommodate Noah, his family, all the animals

God instructed Noah to bring on the ark, and sufficient food, water, bedding, and other supplies needed for 375 days' survival on the ark, plus a few days thereafter for their reestablishment on dry land.

It's worth noting that unlike the boats described in nonbiblical flood stories (see chapter 9), the biblical ark's shape resembles that of large seafaring ships of today. The ratios of its height to width to length have proven optimal for seaworthiness, including flotation stability. One would not expect a man of Noah's time and place to be aware of these optimal ratios without input from a source beyond his own thoughts. Although modern depictions often show Noah's ark to be a rectangular box, Genesis 6 gives only the ark's dimensions. It is silent on its shape. It may or may not have had a curved bow, stern, and hull, the features of modern ships. We simply don't know.

Construction Materials

According to Genesis 6, the ark was either all or largely constructed of wood. The type of wood specified for the ark's construction is "gopher wood." Some English Bibles translate *gopher* as "cypress" wood. However, exactly what kind of tree was harvested for the construction of Noah's ark is unknown. Neither does anyone know all the varieties of trees available for harvesting in and around the Persian Gulf, Mesopotamia, and the Near and Middle East during Noah's time (see chapter 11).

The fact that God specified a certain type of wood for construction of the ark indicates that not just any wood would do. The wood for the ark required certain properties of durability, such as high tensile strength and water repellency, and it would be needed in ample, accessible abundance.

Ancient peoples are known to have built wooden ships roughly similar in size to Noah's ark. For example, in the third century BC, Ptolemy IV, Philopator of Egypt, built the Tessarakonteres, a vessel 280 cubits long, 38 cubits wide, and 48 or 53 cubits high (using the short cubit, 420 feet long, 57 feet wide, and 72 or 80 feet high).[6] This ship was equipped with both sails and oars, as well as the sail hands and oarsmen it required. In this respect, the Tessarakonteres was distinct from Noah's ark. It was designed and built for directed propulsion. This design feature limited Tessarakonteres's size. Since directed propulsion wasn't needed for Noah's ark, no such size limitation applied.

The largest merchant sailing ships of the late eighteenth and early nineteenth centuries were about 300 feet long and constructed of oak. This wood was chosen for its availability at relatively low cost, its tensile strength, and adequate (though barely) water repellency. Shipbuilders soon recognized, however, that constructing oak-timber boats longer than about 300 feet would cease to be economically viable. To keep them free of leaks for journeys of long duration proved too difficult.

For Noah, economics would not have been the concern. Noah's vessel was built for survival, not for making a profit from the sale of cargo. He could afford to build his ark using top-quality, high-strength timbers. Neither was the ark a sailing ship with a specified destination. Noah had no need for directed propulsion, masts, sails, oars, or rudders. His ship simply needed to stay afloat. For this reason, it would demand less tensile strength.

According to a strength index developed by foresters (an average of the Janka hardness index, modulus of elasticity index, modulus of rupture index, and crush index),[7] the strongest species of oak is live oak (*Quercus virginiana*), the species shipbuilders used to construct the largest merchant sailing ships of the late eighteenth and early nineteenth centuries. It has a strength rating of 46.8,[8] above that of all hickory and English walnut species.[9] Compare this rating with that of acacia, which the Bible says was used to build the tabernacle, ark of the covenant, altar of incense, altar of burned offering, and the tabernacle table (Exodus 25:10, 26:26–27, 35:24, 37:1–4, 10, 15, 25–28, and 38:1–7). Gidgee (*Acacia cambagei*), an Australian acacia, has a strength index of 70.7.[10] Snakewood, another type of acacia, has a strength index of 79.8.[11] Nine different tree species, all of them either tropical or subtropical, have a strength index greater than gidgee. Three species have strength indices greater than 90. Ipe, bulletwood, and cumaru, with strength indices of 70.4, 69.8, and 69.8 respectively, are commercially available today in large, usable sizes.[12] All these woods possess excellent water repellency. Thus, it seems reasonable to surmise that the wood used by Noah was as strong (or stronger) and as water repellent as acacia and also readily available to Noah and his work crew.

The wood Noah most likely used would have had a strength index at least 50% higher than that of live oak. Thus, it would have been strong enough for construction of a seaworthy ship 134–158 meters (440–520 feet) long. Its water repellency would have been superior to that of the eighteenth and nineteenth century ships used for international trade. And, to ensure that the ark was as safe as possible, God instructed Noah to seal its hull with pitch (Genesis 6:14). Some debate has arisen over the exact nature of this pitch. Was it a thick tar from petroleum pits? Was it pitch like that exuded by pine trees? Was it some other kind of bituminous material? The answer is insignificant because any one of these three options would have served adequately to waterproof the vessel.

An Impossibly Large Project?

Many skeptics of Genesis 6–9 charge that a ship as large as the one described there, 134 meters (440 feet) long, would have been much too big for a man of Noah's time to construct.[13] They argue that Noah would have lacked the material resources, skills, tools, labor, and time for such a massive and challenging building project—not to mention the intellect and/or imagination. So many facets of the story challenge credulity, but perhaps people of our day and time are the ones who lack imagination.

After all, economics would not have been an issue for Noah. God's command implies provision. From what the Bible reveals about God's character, we can be sure Noah had access to the wealth and resources necessary to follow through with what God asked him to do.

Nowhere does the Bible imply that Noah was asked to build the ark by himself. His family would have helped, and they would likely have hired many workers to assist with the ark's construction. Time would not have been a significant barrier for Noah. While Genesis doesn't state specifically how long the ark was under construction, Genesis 5:32, 7:6, and 7:11 allow for as much as 100 years. A construction effort that continues for several decades can accomplish a sizable project. Those decades also would have given Noah adequate opportunity to call upon his contemporaries to repent and to declare the promise of God's grace for all repentant sinners.

Recently, archaeologists have found evidence that shipbuilding knowledge and capability date back 50,000–44,000 years ago.[14] This early date not only indicates Noah likely possessed such knowledge and capability but that his knowledge and capability was retained to facilitate the great postflood migrations (see chapter 11).

Inadequate Food and Tools for Workers?

A date for Noah's flood earlier than 45,000–40,000 years ago understandably raises questions about whether Noah and his family would have been able to feed all the construction workers needed to complete this massive building project—especially during an ice age. A response to this challenge comes from recent archaeological discoveries, findings that indicate early humans engaged in more sophisticated gathering and preparation of food than was previously considered possible.

A scanning electron microscope analysis of charred plant aggregates recovered from Shanidar Cave in the foothills of the Zagros Mountains along the northeast edge of the Mesopotamian plain revealed the presence of processed pistachios, peas, and mustards.[15] The processing of these items included grinding, soaking, mashing, boiling, and charring. Carbon-14 dates for the aggregates showed that the food had been cooked and eaten at least 40,000 years ago.

Charred plant macrofossils found in Kebara Cave on Mount Carmel, Israel, showed that early humans living 48,000–23,000 years ago were processing and eating wild almonds, acorns, and vetches.[16] In their natural state these foods are filled with cyanide, tannins, and other toxins. To make such food safe for consumption, again, a complex and well-ordered process of grinding, soaking, roasting, mashing, and boiling is required. Evidence of this process indicates that people of that time possessed well-developed tools and the social structure to greatly expand both the quantity and diversity of their food resources.

With the availability and apparent abundance of nutritious food, Noah's generation would have been able to dedicate a significant segment of their population to construction, engineering, and technological development. Genesis 4:22 includes a reference to ancient metallurgy. To date, the only evidence for ice age metallurgy comes from tools cold-forged from iron meteorites. The recovery and accurate dating of metal artifacts from more than 40,000 years ago is bound to be very difficult. Thus, the absence of evidence for such artifacts should not be interpreted as proof that such artifacts did not exist. No scientific reason, therefore, exists to doubt that Noah and his workmen lacked the tools, food, and time to construct the ark.

Inadequate Ventilation?

Some skeptics have seen depictions of the ark showing the presence of just one relatively small window, one cubit by one cubit. They would be right to suggest the inadequacy of this small square to provide essential ventilation for the people and animals on board the ark for any length of time. However, skeptics are mistaken in their understanding of what the text says about the size and shape of the ark's window. Rather than a small square window, an opening ran the entire length of the ark on both sides. In Genesis 6:16 we see that God gave this instruction to Noah: "Make a roof for it [the ark], leaving below the roof an opening one cubit high all around." Such a window under the ark's roof would have provided more than sufficient ventilation while also ensuring that everyone within the ark remained shielded from the torrential rainfall.

Why Take Birds?

Some who question the veracity of the Genesis flood account have asked why Noah would have bothered to take birds onto the ark, given that birds could have rescued themselves simply by flying away to land beyond the reach of the rainfall. Those who question this point of the story overlook the fact that birds can't fly long distances during relentless, torrential downpours. The weight of the rain falling on their bodies overcomes the lift required for flight. And because birds are warm-blooded creatures with exceptionally high body-surface-to-body-weight ratios, they are vulnerable to hypothermia. During Noah's flood, the heavy rain and rising waters would have hindered many bird species from finding a place to survive. Thus, God brought pairs of such species onto the ark. At least some of the bird species on board the ark may have come from outside the vast area covered by the flood, but for these birds to repopulate would have taken far longer had they not been spared.

God wanted Noah and his family to have immediate access to these birds upon exiting the ark. God's obvious intent was that Noah and his family would quickly relaunch and expand civilization after the flood's devastation. This rapid relaunch required the survival of bird species that had formerly inhabited regions covered by Noah's flood and beyond. Several fulfilled significant roles in sustaining ancient agriculture. Others fulfilled critical roles in distributing seeds, fertilizing both terrestrial and aquatic ecosystems, and providing food for wildlife.

Care and Feeding of the Ark's Animals?

Skeptics of Noah's flood often point out that a mere eight people on board Noah's ark couldn't possibly have provided adequate care for the needs of all the animals and birds on board with them. Such skepticism might be valid if, indeed, Noah and his family were rescuing *all* the species of land animals that existed on Earth. However, as explained in chapters 10 and 12, the number of animal species taken on board Noah's ark was limited to the *nephesh* (soulish animals) that had been in relationship with humans.

Nephesh in close contact with people at the time of Noah's flood likely exceeded a few hundred species. The number of animal pairs on Noah's ark could have been as high as a thousand. Caring for a thousand or a few hundred pairs of animals on the ark would have kept Noah and his family busy, for sure. However, with its three decks, the ark offered generous cargo space, including room for animal pens and stalls. Appropriate feed, fodder, and bedding materials could have been stored near the animals. Some animal wastes could have been flushed out of the ark down specially constructed chutes, others combined with food wastes to make fertilizer for future agricultural use.

The ark would have been roomy enough for Noah's family to keep the animals exercised. Storage space would also have been sufficient for tools and materials to be used in postflood construction tasks. It also seems likely that during the multidecade construction of the ark, Noah and his family and their workers installed various labor-saving devices such as dumbwaiters, carts, chutes, rails, and simple plumbing to streamline their work as animal keepers.

Provision for the Future

In addition to the obvious goal of keeping the ark's passengers safe and secure for the duration of the deluge, a secondary purpose for an ark of such great size would have been to carry essential supplies for the rapid relaunch of civilization. These supplies most likely included seeds for planting crops and seedlings for planting groves. Genesis 9:20 mentions that Noah planted and tended a vineyard soon after the flood. God's clear desire to see civilization restored implies that Noah and his family brought with them whatever they needed to do so. (The matter of how eight people could give rise to the present genetic diversity in the human population is addressed in chapter 16.)

The "gopher wood" waterproofed with "pitch" would have provided excellent materials for the construction of dwellings, sheds, carts, etc. So, we see that the dismantled ark itself would have provided much of what Noah and his family needed for launching civilization.

The ark gave future generations of humanity a reminder and reassurance that God always delivers from judgment all who repent of their rebellion against their Creator. Again, another reason for the ark's size is to demonstrate that God will make ample room for *anyone* and *everyone* who repents.

An Uphill Journey?

One more objection raised by global flood proponents to any suggestion of a flood less than globally expansive is the report of the ark's landing on the mountains of Ararat rather than floating far downstream, either into the Persian Gulf or out into the Indian Ocean. Genesis 8 provides a helpful response. God "sent a wind over the earth and the waters," a wind so strong and persistent that it played a major role in evaporating the waters of the flood.

Physicist Alan Hill used computer models to simulate various scenarios whereby wind would push Noah's ark against the current of outflowing water and up against the foothills of Ararat.[17] He calculated the drag forces from outflowing water on the part of the ark below the water line and the wind forces on the part above the water line that would tend to drive the ark upstream.

Hill showed that a wind of 80 km/hour (50 miles/hour) average velocity with gusts up to 88–97 km/hour (55–60 miles/hour) could push a fully loaded ark weighing 4,550 tonnes (5,000 tons) some 1,100 kilometers (700 miles) inland, where it would run aground on one of the hills/mountains of Ararat in as few as 40 days. If the fully loaded ark weighed significantly more than this estimated tonnage, the wind velocity required to drive it there would have been greater, of course. However, Hill demonstrated that in all cases, the required wind velocities fell well within the realistic range for a large, stalled cyclonic storm over the Persian Gulf and Mesopotamian region.

Searching for Noah's Ark

Over the past several decades, many people have claimed to discover the wooden remains of Noah's ark near the summit of Mount Ararat.[18] A boat-shaped geological structure (the Durupinar formation) 164 meters (538 feet) in length at 2,000 meters (6,500 feet) elevation on the south side of Mount Ararat has been said by others to be the petrified remains of Noah's ark.[19] Scholars have disputed both these claims, with good reason. Archaeologists and geologists have been consistent in concluding that the boat-shaped structure is a natural formation, not a petrified ship. Hikers and climbers frequently see similar geological rock structures that approximate the size and shape of Noah's ark on their alpine hikes in the mountain ranges of North and South America and Europe.

Mount Ararat is a glaciated peak. Glaciated peaks tend to be unsafe for climbing at certain slope angles during certain parts of the day. For thousands of years, ardent mountain climbers have dealt with this danger by building small huts in which to camp overnight just below the treacherous zone. They make their ascent to the summit early in the morning and return to the quickly constructed hut before the dangerous daylight hours. Often these huts are left behind as a courtesy to future climbers. Most likely by far, the pieces of wood found on Mount Ararat are the remnants of ancient climbing huts, not pieces of Noah's ark.

Reasons for skepticism include these three, at least. First, the wood fragments found thus far lack the necessary strength to have been structural components of Noah's ark. Second, these wood fragments date too young to be fragments of the ark. Third, Genesis 8:4 says that the ark came to rest on the *harim* of Ararat. As explained in chapter 12, the hills and mountains of Ararat encompass an area of about 250,000 square kilometers (about 100,000 square miles). A more reasonable location to search for the landing place of Noah's ark would be the low foothills of Ararat, those that adjoin the Mesopotamian plain (see the map in figure 12.1 in chapter 12).

Not even the low foothills of Ararat adjoining the Mesopotamian plain, however, represent a reasonable locale in which to find the ark's remains. Its wood would have been much too valuable for the postflood people to leave lying on the ground. They would have quickly exploited the wood for their own building needs. According to Genesis 11, postflood construction projects were grandiose. For this reason, the best place to search for the remains of Noah's ark may be in or near the ruins of the ancient cities named in Genesis 11. This search would be no easy task, however, since all these cities have either been burned to the ground or utterly destroyed. Archaeologists have yet to discover the ruins of any city or village that predates 12,000 years ago.

The opening chapter of this book comments on widespread reaction to the Genesis account of Noah's flood, the fact that many people point to this story—or at least to their understanding of it—as an excuse for rejecting the Bible as the Word of God. As noted in chapter 8, several leading Christian theologians agree that the biblical account of Noah's flood includes so many "fantastic" elements that it must be accepted as hyperbole or myth or "mytho-history," not as literally and historically true. Any attempt to interpret the account as a plausible historical narrative, as I do, is to invite criticism and/or serious misunderstanding. The aim of the chapter that follows is to address various challenges to this interpretive model or to others like it.

Chapter 15

"Fantastic" or Factual?

As chapter 8 reveals, a growing number of Christian theologians are now inclined to label the stories in Genesis 1–11 as something *other* than historical narrative. While interpretations of these passages may vary from person to person, they tend to agree on one point: these early chapters of Genesis, including the account of Noah's flood, must be understood and embraced as containing "fantastic" components,[1] to use the term coined by William Lane Craig. The stories of creation, the fall, the flood, and Babel, all include, they would say, elements "which, if taken literally, are so extraordinary as to be palpably false."[2]

Some of these theologians deny the miraculous features in Genesis 1–11. Others, like Craig, do not. Craig clarified that the fantastic elements in Genesis 1–11 "concern nonmiraculous features of the story that, if taken literally, are palpably false."[3]

This response to Scripture represents a clear departure from the historic doctrine of biblical inerrancy—an unnecessary departure, at that. Often, the "implausible" elements, when carefully considered, prove to be misinterpreted or misunderstood elements. The following paragraphs serve to illustrate this point by addressing each item in Craig's list of "fantastic" elements that is relevant to Noah's flood: vegetarianism, lifespans beyond 120 years, the vast extent of the flood, the features of the ark, the Table of Nations, and the Tower of Babel.[4]

Fantastic Vegetarianism?

Only *if* a reader of Genesis assumes from the message of chapter 9, verse 3, that *neither* humans *nor* other macrofauna (animals visible to the naked eye) were consuming any meat until after Noah's flood,[5] a strictly vegetarian diet for all would, indeed, become an implausible element of the text. However, this is not what Genesis 9:3 says or implies. After the flood, God removed a dietary restriction from humans alone and the reason for this change becomes clear in the verses to follow. God's commands are not arbitrary.

Carnivorous animals play a part in God's good design for stable, healthy ecosystems. They predominantly prey upon weak, diseased, or genetically impaired animals. By removing these individuals from herbivore flocks, carnivores alleviate suffering caused by disease and genetic issues. By limiting herbivore population levels, carnivores prevent overgrazing and, thereby, protect the food supply of other creatures, including humans. The carnivorous animals God created help maximize herbivores' health and well-being and minimize their suffering and death rate. In fact, carnivores enhance entire ecosystems, as the reintroduction of grizzly bears and wolves into Yellowstone National Park recently demonstrated.[6]

Genesis 9:3 need not, I would say *cannot*, be rightly interpreted to include a reference to all Earth's animals. When God said, "Just as I gave you the green plants, I now give you everything," he was speaking specifically to Noah, as becomes clear in Genesis 9:4. God's command, in this verse, is a prohibition against eating "meat that has its lifeblood still in it." Humans, not animals, are able to drain blood from their prey before consuming it. Nonhuman carnivores either cannot or do not, given the competition they face from other predators.

In Genesis 9:5–6, God adds a prohibition against murder, which he defines as shedding "human blood." Again, the context is specific to humans and exclusive of nonhuman animals. These are God's instructions to and for people. The words of Genesis 1:24–25 imply the existence of carnivores from early on. This passage tells us God created two distinct kinds of large-bodied mammals, "livestock" and "wild animals" (*behema* and *hayya*, respectively, in the Hebrew text). The Reformation Study Bible includes this comment on Genesis 1:24–25: "The contrast between wild and domesticated animals differentiates carnivores from cattle."[7]

Passages such as Job 38:39–41 and Psalm 104:21, 24–28 tell us that carnivores receive their prey from God's hand. Psalm 104:27–28 says, "All creatures look to you [God] to give them their food at the proper time. When you give it to them, they gather it up; when you open your hand, they are satisfied with good things."

To support his claim that at the time of Adam and Eve *all* animals subsisted exclusively on "green plants," Craig cites the commentary on Genesis 1:29–30 by Old Testament scholar Victor Hamilton.[8] However, both these scholars overlook an important contextual distinction between Genesis 1:29 and Genesis 1:30. In Genesis 1:29, God decrees that the human diet should consist of "seed-bearing plants" and the "fruit" and "seeds" from trees. Here, God orders a vegetarian diet for humans. In Genesis 1:30, God explains that "green plants" will be the primary food source for all terrestrial creatures.

The Hebrew words for green plants in this text are *yereq* and *'eseb*, respectively. The *Theological Wordbook of the Old Testament* defines *yereq* as "green or greenness"[9] and *'eseb* as "grass, herb, vegetation, or plants."[10] The passage seems to emphasize the importance of what scientists today would refer to as photosynthetic vegetation. In Genesis 1:29–30, God is instructing humanity on how to manage Earth and all its resources, both living and nonliving, for the benefit of all Earth's life, including their own.

Green plants are the foundation of the entire food chain. Without green plants, no animal on Earth gets to eat. It seems apparent that God emphasized to Adam and Eve (and, thus, to all their progeny) an important aspect of their job. Because all life fundamentally depends upon green plants for survival, the wise management of photosynthetic vegetation is a crucial priority. I see no compelling reason to conclude that all animals living before Noah's flood ate nothing but plants. In this case, vegetarianism does not constitute a fantastic element.

Fantastically Long Lifespans?

Based on many centuries of observation, we know that any human being, even with the most advanced medical health technology, can't live 900+ years. Nevertheless, Genesis 5:5, 5:27, and 9:28 declare that Methuselah, Adam, Noah, and others *did* live that long. To explain how these statements about the longevity of early humans could be plausible, rather than "fantastic," seems a daunting challenge. Many have tried and failed.

Today, the radiation environment to which all humans are exposed limits the human lifespan to a maximum of about 120 years, as stated in Genesis 6:3. Constant exposure to three sources of deadly radiation is the major limiting factor:

1. short-wavelength ultraviolet radiation from the Sun,
2. cosmic radiation from supernova remnants, novae, black holes, and gamma-ray sources, and
3. decay of radioisotopes in Earth's soil, surface rocks, and crust.

The key question, then, is whether humans' exposure to these forms of radiation could have been different from levels of exposure since that very early era. Exposure to the first source, solar radiation, mitigated to some degree by clothing, was less of a concern when Earth's stratospheric ozone shield provided greater protection from solar short-wavelength ultraviolet radiation than it does today.[11] Exposure to solar radiation has become more significant since the weakening of that shield, but solar radiation is a relatively minor factor in limiting longevity. Cosmic radiation, on the other hand, is a major factor, and that's where a more significant change has occurred.

Analyses performed over a period of three decades by two astronomers, Anatoly Erlykin and Sir Arnold Wolfendale, show that most of the deadly cosmic rays impacting Earth today come from a single supernova and its remnant (see sidebar Cosmic Radiation Event).

The relatively recent supernova event identified by Erlykin and Wolfendale may provide some help in explaining the longer maximum lifespans of people living before Noah's flood. Humans living before the event would have been exposed to much less of the deadly high-energy, heavy-nuclei cosmic radiation than we experience today. For this reason, they likely experienced a much lower risk of cancer, cell damage, and organ failure. If the supernova identified by Erlykin and Wolfendale erupted near the time of Noah's flood, people living before the flood may have been able to live considerably longer than people born centuries after the event.

For a lifespan of 900+ years to be possible, however, more than merely the avoidance of cosmic and solar radiation would be essential. Humans are also impacted by terrestrial radiation sources. These sources, including radon, radium, thorium, and uranium radioisotopes, are pervasive in Earth's igneous rocks and coal deposits. To avoid exposure to igneous rocks is all but impossible today.

Concrete structures and paved or gravel roads are everywhere. Modern life requires people to live in and/or travel through regions of igneous rock, such as mountains, hills, and gravel-lined valleys. At the time of Noah, however, the people likely spent all, or nearly all, their time in sedimentary plains, and coal was not their fuel source. Thus, they would have been protected, at least to a large degree, from these deadly terrestrial radiation sources.

The possibility to live significantly longer entails obedience to God's dietary instructions to Adam and Eve. By comparison to vegetable matter, meat contains a far greater concentration of heavy metals—up to a thousand times greater. These metals include mercury, lead, thorium, and uranium. While a few plant species under certain conditions may also concentrate metals to dangerous levels,[12] it's

Cosmic Radiation Event

In 1997, astronomers Anatoly Erlykin and Arnold Wolfendale noted two prominent peaks in the high-energy end of the cosmic ray air shower spectrum: one identified with oxygen nuclei at ~3 PeV, the other with iron nuclei at ~12 PeV.[13] In 2005, they noted the "hint" of a third peak at 50 PeV.[14] In 2008, the Tibet Air-Shower Collaboration indisputably affirmed the high-energy peaks or "knee" in the cosmic ray air shower spectrum.[15] This cosmic ray spectral profile bears the signature of a single radiation source: a recent, local supernova eruption.[16] This same signature also appears in the cosmic ray proton, helium, lithium, positron, and antiproton spectra.[17] The observed very-high-energy knee implies that the deadliest cosmic radiation incident on Earth comes from a nearby, recent supernova eruption. It accelerated oxygen and iron nuclei and possibly even heavier nuclei to extremely high velocities via shocks in the hot interstellar medium. Erlykin and Wolfendale concluded that the supernova was so recent and so nearby that our solar system likely resides just inside the shell of its expanding remnant.[18] Erlykin and Wolfendale calculated the date when deadly radiation from the supernova eruption first arrived on Earth: between 30,000 and 100,000 years ago.[19]

broadly true that the higher a creature is on the food chain, the higher the concentration of deadly metals and pollutants in that creature's body.

The accumulation of such metals in the human body through consistent consumption of meat would make lifespans of several hundred years impossible. However, the combination of a strict vegetarian diet with lower exposure to radiation could potentially extend the human lifespan to a significant degree, but to the extent of 900+ years is not yet confirmed.

Today, our chromosomes shorten as we age, and the rate at which this shortening occurs appears optimized to protect us from developing malignant tumors. This telomerase activity, as it is called, is the primary reason no one today lives past about 120 years. However, where exposure to deadly radiation is significantly less, the need for aggressive telomerase activity diminishes. That is, people living before Noah's flood may have been genetically endowed with a level of telomerase activity much lower than that of people living today. Their load of accumulated deleterious mutations would also have been lower.

The early chapters of Genesis strongly suggest that although people had the potential to live 900+ years, their average lifespan may have been as low or lower than the average lifespan today. As noted in chapters 2 and 10, it's likely that most people living before Noah's flood died from unnatural causes, including murder. Further, given the spiritual state of Noah's contemporaries, described in Genesis 6, it seems highly unlikely that they obeyed God's dietary instructions.

Why did God grant preflood peoples a maximum lifespan of 969 years and limit us to only 120 years? Perhaps he did so to underscore the lesson explained in chapter 10 that, given the nature of human sin, we are far better off with a 120-year maximum lifespan.

Fantastic on What Other Basis?

Many Christian scholars agree that Noah's flood could not have been global in extent. Craig has commented favorably on my previously published "brief but crushing summary of the [geological] difficulties" with the claim of a globally extensive flood.[20] (Chapter 13 provides an updated and lengthier summary.) Nevertheless, he and others remain convinced that the Genesis text demands a global interpretation, which makes it "fantastic" as a whole.

Craig and others agree with young-earth creationist and global flood proponent Jonathan Sarfati, among others, that reference to "all the mountains under all the heavens" (Genesis 7:19) requires interpreting "all" in the global sense, despite contextual clues to the contrary.[21] Appeals to texts outside of Genesis indicating that Noah's flood was less than globally extensive are, for unknown reasons, simply dismissed as "irrelevant."[22] Thus, chapter 12 of this book is devoted to showing that a literal, historical interpretation of Noah's flood based on the Genesis text alone is possible. Though the flood's devastation was universal with respect to humanity, it need *not* have been globally extensive to fulfill God's purpose for it.

What may have stirred a major reaction to my less-than-global interpretation of Noah's flood is the faulty label attached by some scholars: "local." This label fails to acknowledge that in my proposed model for the flood, its waters reached far beyond the Persian Gulf region.[23] As chapter 13 explains, the likely timing of Noah's flood—during the era when the global mean temperature was fluctuating widely and rapidly—means that human settlement likely ebbed and flowed with the advance and retreat of massive ice sheets. Humanity at the time of Noah's flood may have retreated all the way into the Persian Gulf area, southern and central Mesopotamia, the Gulf of Oman, and lowland regions bordering the Gulf of Oman. But the area occupied by humans may also have extended to all the lowland regions of eastern Africa, the Near and Middle East, and the land underneath what is now the Red Sea and the eastern Mediterranean Sea. If so, that whole area would have been flooded.

Craig and other theologians estimate a date for Noah's flood more recent than 10,000 years ago. I do agree that a date so recent would make the account of Noah's flood fantastic and implausible. It's far more probable, biblically and scientifically, that Noah's flood occurred previous to 45,000 years ago (see chapter 11). Such an early date would remove the implausibility factor.

Identification of the animals impacted by Noah's flood has led to another misunderstanding. Craig and others interpret Genesis 6–9 as implying that all Earth's land animals were killed in the flood.[24] If that were the case, again, I would agree that Noah's flood is a fantastic story. However, from careful study of the Hebrew words used for animals in Genesis 6–8 and thoughtful consideration of which animals are damaged by exposure to human sin, we can see that only *nephesh* (soulish animals) were destroyed in the flood—and only those owned or closely associated with Noah's wicked contemporaries. Again, a global flood is not necessary, and animal death need not render the story "fantastic."

Correct identification of the animals also impacts the plausibility of the ark and its capacity. Even if built according to the longest cubit, the ark would have been too small to accommodate a pair of every species of Earth's land animals. Further, Noah's family would have been incapable of caring for all these animals for an entire year. Either the ark's animal passengers or the ark itself would, by such an interpretation, be considered fantastic. However, as chapter 14 shows, a literal, historical interpretation of Genesis 6–9 makes no such fantastic claim.

Fantastic Table of Nations?

Many scholars claim that to believe the 8.05 billion people on Earth today[25] owe their descent to the eight people on board Noah's ark is unreasonable. According to Craig and at least one well-known theologian of the last century, Hermann Gunkel, the idea that the nations of the earth originated from a single family is worthy of ridicule.[26] I would agree with Craig, Gunkel, and others *if* Noah's flood had occurred less than 10,000 years ago. However, an abundance of biblical and scientific evidence indicates it did not. Rather, it's likely to have occurred between 120,000 and 45,000 years ago.

If the flood occurred between 120,000 and 45,000 years ago, no genetic refutation would nullify the conclusion that all the world's present population descended from the eight people on Noah's ark. This claim is substantiated in chapter 16. The Table of Nations in Genesis 10, therefore, need not be treated as a fantastic element.

Fantastic Diversification of Language?

If, in fact, the only human survivors of the flood were Noah and his family members, just eight people in all, we can reasonably infer that all the cultures and languages of the world trace back to the culture and language of that family. Thus, scholars who claim a recent date for the flood are compelled to assign an even more recent date for the biblical Tower of Babel story. According to Genesis 11:5–9, the Babel incident marks the beginning of language diversity and humanity's spread to all accessible regions of the globe.

The story of Babel, like the story of Noah's flood, is considered implausible by a growing number of Christian scholars today. Craig comments, "As a result of modern linguistics studies, we know that such a sudden, unitary origin of the world's languages is palpably false."[27] In other words, the Babel account, too, must be considered "fantastic." However, the implausibility stems from the scholars' assumed timing of the flood and Babel, more than from these events themselves.

If the Tower of Babel incident had occurred as recently as (or more recently than) the Neolithic revolution, circa 12,000–11,000 years ago, its plausibility would, indeed, be an issue. However, the Bible doesn't assign a recent date for the Tower of Babel any more than it does for the flood. Here, again, the scientific record provides a helpful estimate of its timing.

According to Genesis 11:5–9, the diversification of human language coincided with the scattering of people to all regions of the earth. As mentioned in chapter 11, the great migration of humanity from eastern Africa and the Near and Middle East into Europe, East Asia, and Australia took place some 50,000–40,000 years ago. This timing coincides with that of language diversification (see chapter 11). Genesis 11 makes no claim as to how long the scattering of people over all the earth took, nor does it specify how long was required for a single human language to become many languages. Nothing in the biblical text suggests "suddenness."

The scientific record shows that human migration and settlement into Europe, East Asia, and Australia occurred roughly between 50,000 and 40,000 years ago and into North and South America, roughly 16,500–14,000 years ago. These dates are consistent with linguists' estimated dates for the diversification of human language (again, see chapter 11 documentation).

Plausible After All

Craig deserves both praise and appreciation for his book *In Quest of the Historical Adam.*[28] Not only does the book provide a thoroughly referenced list of fantastic elements, but it demonstrates scholarly humility. In it, Craig acknowledges, first, that if Genesis 6–11 is filled with fantastic elements, then to interpret the text literally and historically would be difficult, if not impossible. Because he sees these fantastic elements, he must conclude that Genesis 6–11 is mytho-history: partly historical, mostly mythical. However, he also acknowledges that if Genesis 6–11 is free of fantastic elements, then one can reasonably interpret the whole of this much-debated text literally and historically. In his book, *Set Apart*, historian Joseph R. Miller offers comprehensive historical and exegetical evidence for the absence of fantastic elements in Genesis 1–11.[29]

One major point of controversy over the historicity of Noah's flood remains to be addressed. In the context of Noah's flood, this element has two distinct components: the evolution of animals and the evolution of humans. The next chapter focuses on what can and can't be attributed to natural evolutionary processes.

Chapter 16

Life Before and After the Flood

The issue of biological evolution looms large in any consideration of how to interpret the Genesis account of Noah's flood. Those who claim the flood was globally extensive and deadly to all Earth's creatures, other than those on the ark, must explain the existence not only of the people but also of the plants and animals on Earth today. Did all these life-forms descend from the ark's passengers—especially in the brief time span proposed? Even Earth's marine life and plants must be accounted for. The sea creatures and vegetation that existed before the flood would have been obliterated by the extreme lateral and vertical tectonic movements that global flood proponents assert occurred during Noah's flood (see chapter 13).

Evolution or Diversification?

Ecologists have determined that the total number of eukaryotic species on Earth today adds up to about 8.7 million species.[1] Eukaryotes are all organisms whose cells possess a membrane-bound nucleus, which includes plants, animals, and fungi. Even many unicellular species are also eukaryotic. Of these 8.7 million eukaryotes, 6.5 million are terrestrial species and the remaining 2.2 million are marine species.[2]

Global flood proponents do acknowledge that Noah's ark could not have accommodated 8.7 million species and that the eight people on board Noah's ark could not have kept them all alive for a year. In response to this obvious impossibility, they propose that Noah and his family harbored only pairs of animals categorized on standard taxonomy charts above the species and genus levels (see sidebar Taxonomic Rank). Most global flood advocates also assert that only terrestrial animal pairs were taken onto the ark. They presume that by

Taxonomic Rank

Taxonomic rank is an organizing system in biology where characteristics among organisms are ranked according to how many characteristics the organisms at each rank level share in common. For species (the lowest rank), the common characteristics are the most numerous whereas for domain (the highest rank), the common characteristics are the least numerous.

A helpful example would be the red fox. It is the only member of the *Vulpes, vulpes* species. The genus *Vulpes* includes all the true foxes. The family Canidae includes foxes, wolves, coyotes, dogs, and jackals. The order Carnivora includes all carnivorous mammals: canines, cats, hyenas, bears, seals, weasels, raccoons, skunks, and mongooses. The class mammalia includes all hairy, warm-blooded animals who nurse their young. The order Chordata includes all vertebrate animals plus all invertebrate animals with notochords (a long, flexible, midline chord of nerves that runs along most of the length of the animal). The kingdom Animalia includes all animals. The domain Eukarya includes all organisms whose cells possess a nucleus.

some unspecified means, marine plant and animal species survived the devastations they insist occurred.

In their book *The Genesis Flood*, John Whitcomb and Henry Morris estimate the ark may have carried as many as 17,500 pairs of animals,[3] but only if essentially all these creatures hibernated for the entire duration of the flood. Geologist John Woodmorappe calculated that the maximum conceivable carrying capacity of Noah's ark would have been 8,000 pairs of animals.[4] His calculation assumed the animals would be as densely packed as animals in research laboratories or factory farms. However, biologists have questioned whether animals so densely packed for so long could have survived, much less retained their ability to reproduce.[5] Answers in Genesis biologist Nathaniel Jeanson proposes that only pairs of each terrestrial family were brought onto Noah's ark, a number totaling just 1,700 pairs.[6]

In the global flood proponents' models, God does not create millions of species of life. Rather, he creates a few thousand kinds from which millions of species have diversified. The few thousand kinds on board Noah's ark, they say, diversified at an exponentially rapid rate to produce millions of species within a few centuries or less. As Jeanson wrote, "Once the animals stepped off the Ark, their reservoir of DNA differences could have easily translated into a massive amount of morphological change."[7] However, how or why the diversification rate dramatically slowed within just a few centuries or why there is no physical evidence for the purported slowing is left unexplained.

Global flood proponents claim, for example, that zebras, horses, donkeys, onagers, and several other horse-like creatures quickly "diversified" from a single horse-like pair on board Noah's ark.[8] The entire cat family—tigers, lions, leopards, cheetahs, jaguars, panthers, ocelots, caracals, servals, bobcats, and more, including the ancestors of housecats—rapidly diversified, they say, from a single feline pair on the ark.[9] To support their claims, they cite hybridization experiments in which lions have been crossed with tigers to produce ligers, zebras with donkeys to make zonkeys, whales with dolphins to make wholphins, and goats with sheep to make geep. It must be noted, however, that these hybridization experiments are irrelevant. They require human intervention and the resultant offspring do not reproduce after their own "new" hybrid kinds.

In the wild, animals prefer to mate with members of their own species. While hybridization has occurred in the wild, it is limited to genetically similar species, such as Alaskan brown bears and polar bears, and takes place only when climate change episodes force the overlap of their habitats.[10]

Proponents of this diversification process assiduously avoid using the term *evolution*. They do use such terms as "common descent," "microevolution," and/or "adaptation," but they strongly reject the universal common ancestry of all species and macroevolution.

Many evolutionary biologists have pointed out that the attempt by global flood proponents to distinguish diversification of species from the evolution of species is disingenuous, at best, and fails spectacularly in terms of science.[11] Philip Kitcher, Columbia University philosopher of science, wrote that the rates of speciation "to manage the supposed level of species diversification are truly breathtaking, orders of magnitude greater than any that have been dreamed of in evolutionary theory."[12] In other words, global flood advocates are forced by their position to subscribe to ultraefficient and extremely rapid, naturally occurring biological evolution.

The Biblical "Kind"

According to Genesis 1, God made Earth's creatures "according to their kinds."

Global flood proponents argue that the biblical "kind" equates with the taxonomic level of order, family, or subfamily. The biblical Hebrew word translated "kind" is *mîn*. As this word is used in Genesis 1, 6, and 7, Leviticus 11 and 19, and Deuteronomy 14, it indicates that animals belonging to a *mîn* reproduce the characteristic features of that *mîn* from one generation to succeeding generations.[13]

Global flood proponents argue that the definition of *mîn* is broad enough that a male and a female animal belonging to a *mîn*, after Noah's flood, could reproduce to yield all the species within a taxonomic family. (The number of species per animal family runs from 80 to 50,000.) Some global flood proponents assert that the definition of *mîn* is sufficiently broad that some herbivores could have diversified after Noah's (relatively recent) flood to become carnivores.[14]

Such a broad definition of *mîn* appears to contradict the usage of the word in the only Old Testament passages that name specific animals as belonging to separate *mîn*. Leviticus 11:22 lists the locust, katydid, cricket, and grasshopper as distinct *mîn*. Deuteronomy 14:12–18 identifies the eagle, falcon, hawk, red kite, black kite, vulture, and black vulture as separate *mîn*. In both Leviticus 11:15–17 and Deuteronomy 14:15–17, the horned owl, screech owl, little owl, great owl, and white owl are referred to as separate *mîn*.

With respect to birds and mammals, the definition of *mîn*, based on these biblical passages, appears no broader than that of the English word *species*. With respect to insects, the definition may be as broad as the English term *genus*, but no broader. Careful consideration of both the biblical and scientific data shows that the rapid, aggressive postflood multiplication and diversification of the animal pairs proposed by global flood proponents lacks any conceivable credibility.

Postflood Human Genetic Diversity

A literal, historical interpretation of Genesis 6–11 implies that today's human population traces its lineage all the way back to the eight people of Noah's family. But how feasible is it that the genetic diversity observed in the human population of our time could have derived from just these eight people? Is this claim as implausible as the global flood claim about the animals?

The early date for Noah's flood certainly helps. As noted in chapter 11, Noah's flood probably occurred before 45,000 years ago. It could have occurred as far back as 120,000 years ago. Several tens of thousands of years of human reproduction allows for substantial genetic diversity.

Geneticists use three different mathematical models to determine how much time was required for development of the genetic diversity observed in the current human population. These three models are based on (1) estimated mutation rates, (2) frequency of linkage disequilibrium, and (3) frequency of incomplete lineage sorting. Linkage disequilibrium refers to the nonrandom association of gene variants at different positions along a chromosome. It is affected by natural selection, mutations, genetic drift, genetic recombination, distribution and structure of the population, and mating habit contraints.

Incomplete lineage sorting refers to ancestral gene copies' failure (looking back over time) to coalesce into a single common ancestral copy. That is, it refers to the observation that the most related subpopulations do not always inherit the most related genes. Incomplete lineage sorting may be caused by nonrandom mutations, major ecosystem disturbances, gene flow from disruptions or anomalies in mating practices, and horizontal gene transfer.

The three mathematical models predict that for an initial population size as low as eight individuals, 100,000 years or more would be required to produce the genetic diversity observed in the human population today. However, the predicted traceback times show discordance, which isn't surprising since all three models are based on simplified, idealized mathematical relationships. These relationships either fail to consider all the relevant variables or, in several cases, crudely estimate them.

For example, variations in population dynamics resulting from geography, migration, climate change, culture, and technology are either assumed to be irrelevant or simply guesstimated. Further, geneticists can't accurately determine mutation rates earlier than 45,000 years ago, nor can they consistently determine the locations of mutations along the chromosomes throughout human history. As population geneticist Erika Hagelberg says, "It is becoming apparent that mtDNA evolution is more complex than previously believed. Anomalous mutation patterns perturb phylogenetic assumptions."[15] In other words, poorly understood mutation patterns disturb the basic ideas about how genetic diversity develops within the human species.

Mitochondrial DNA (mtDNA) variants trace humanity's maternal lineage, given that all humans get their mitochondrial DNA exclusively from their mothers. Y-chromosomal variants trace humanity's paternal lineage since, with rare exceptions, only males possess a Y chromosome. (Women with Swyer syndrome have a Y chromosome, female genitalia, and internal reproductive organs but lack eggs and, therefore, cannot pass on their Y chromosome.)

Four research teams used different combinations of the three mathematical models to calculate the time back to a single ancestral mitochondrial DNA sequence (mitochondrial Eve). They produced the following dates to which the present human mitochondrial DNA diversity could be traced back to an individual female:

Phillip Endicott et al.:	108,000 years ago[16]
Qiaomei Fu et al.:	157,000 ± 39,000 years ago[17]
Adrien Rieux et al.:	143,000 ± 34,000 years ago[18]
G. David Poznik et al.:	123,000 ± 25,000 years ago[19]

When the models are used to determine the time back to a single ancestral Y-chromosome sequence (Y-chromosome Adam), they yield the following dates to which the present human Y-chromosome diversity could be traced back to an individual male:

Wei Wei et al.:	108,000 ± 7,000 years ago[20]
Fulvio Cruciani et al.:	141,500 ± 15,600 years ago[21]
G. David Poznik et al.:	138,000 ± 18,000 years ago[22]
Rosaria Scozzari et al.:	163,000 ± 41,000 years ago[23]

It's worth noting that the authors of these research studies describe both their dates and error bars as "estimates," not measurements. As previously mentioned, several systematic effects remain unknown—or as yet undetermined. Nevertheless, the maximum range of dates for Noah's flood (see chapter 11) falls within the range of these roughly determined dates for mitochondrial Eve and Y-chromosome Adam.

Y-Chromosome Noah

The four women on board Noah's ark likely are unrelated, biologically. Therefore, the mitochondrial DNA date back to Eve could be much earlier than the date estimates for these four women. Such is not the case for the four men on Noah's ark. Noah's three sons all received their Y-chromosome DNA from their father, who received his from his father, and so on. That is, the estimated date for the Y-chromosome Noah would be the same as the date for the Y-chromosome Adam.

Researchers offer several reasons, however, to doubt the earliest dates for the Y-chromosome Adam/Y-chromosome Noah. A team of over 40 participants in the 1000 Genomes Project consortium constructed a calibrated phylogenetic tree (how living things are connected). This tree "shows bursts of extreme expansion in male numbers that have occurred independently among each of the five continental superpopulations examined, at times of known migrations and technological innovations."[24]

Astrophysicist and honorary research fellow in genetics Iain McDonald advised research geneticists attempting to date human ancestry through analysis of Y-chromosome data "to approach their data with an abundance of caution."[25] He continued, "the small numbers of mutations involved in many of these calculations mean a single mutation can make a significant difference to the TMRCAs [time to most recent common ancestor] involved, thus whether mutations are counted correctly is as important a factor in obtaining accurate TMRCAs as whether the correct method is used."[26]

Similarly, known reasons exist to doubt the antiquity of date estimates for mitochondrial Eve. Geneticists Eleanor Miller, Andrea Manica, and William Arnos have noted that "modern populations will often carry ghost signals of demographic events that occurred far away from their current location."[27] Concerning mitochondrial DNA mutation-rate estimates, geneticists Sanni Översti and Jukka U. Palo have written, "In general, it is extremely difficult to imagine that bias caused by sequencing errors and/or DNA degradation are affecting only one subhaplogroup."[28]

As with calculated dates for Y-chromosome Adam/Y-chromosome Noah, dates for mitochondrial Eve are dependent on presumed times of natural evolutionary divergence between hominids/chimpanzees and anatomically modern humans and between Neanderthals and anatomically modern humans. That is, these calculations assume anatomically modern humans and Neanderthals share a common evolutionary ancestor and humans and chimpanzees share a common evolutionary ancestor.[29] As geneticists Översti and U. Palo have demonstrated, "a clear positive correlation exists between the age of the calibration point [assumed time back to the common evolutionary ancestor] and the estimated age of the MRCA [most recent common ancestor]."[30] If one acknowledges that anatomically modern humans are created in the image of God (Genesis 1:26–27), rather than the product of common evolutionary descent, then the calculated dates previously listed are likely incorrect.

Recent Supernova Implications

In a study of human DNA mutation rates, a team of four geneticists discovered "a rapid decay in evolutionary mutation rate between approximately 2,500 and 50,000 years ago and a slow decay from 50,000 to 6Mya."[31] Astronomers

Anatoly Erlykin and Sir Arnold Wolfendale provide an explanation, as mentioned in the previous chapter. They observed an anomaly known as the "knee" feature in the very-high-energy region of the cosmic ray spectrum. This feature is best explained by a single supernova explosion, relatively recent and relatively close to our solar system.[32] An independent team of five astronomers estimated that this supernova erupted 70,000–100,000 years ago at a distance of 938 ± 100 light-years from Earth.[33] Later, a different team of three astronomers using spectral analysis determined a date of roughly 68,000 years ago.[34] In a separate study, Erlykin and Wolfendale, along with physicists Evgeny Berezhko and Leonid Ksenofontov, determined that the majority of the highest energy, heavy-nuclei cosmic rays (the ones most likely to cause mutations in human DNA) showering Earth today come from this particular supernova.[35]

If the supernova identified by Erlykin and Wolfendale erupted during the time of Noah, it could have produced enough mutations in human DNA to explain the present-day diversity in human DNA even if Noah lived significantly more recently than 100,000 years ago. It could also yield more recent dates for mitochondrial Eve and Y-chromosome Adam.

Conservation Biology Implications

When mathematical models for propagation of genetic diversity are tested by conservation biologists in their field experiments, these models consistently and substantially underestimate the propagation of genetic diversity. To date, researchers' field studies have focused on Mouflon sheep,[36] gray whales,[37] blue whales,[38] Przewalski's horses,[39] copper redhorses,[40] common toads,[41] white-tailed deer,[42] and white-tailed eagles,[43] all with known initial and final population sizes. The studies' goal: to determine initial and final genetic diversity.

In every one of the eight experiments, the development of genetic diversity far exceeded what the mathematical models had predicted. The experiments showed, further, that the models' underestimation of genetic diversity development correlated closely with the species' generation time.[44] The longer the average time between birth and reproductive maturity, the greater the propagation of genetic diversity.

The generation time (years between birth and giving birth) of anatomically modern humans is extraordinarily long. While generation times may vary significantly among diverse human populations, population geneticists have determined that over the history of anatomically modern humans, the average generation time for males equals 30.7 years and for females, 23.2 years.[45] These generation times are much longer than those of other known creatures, with the exception of some whale and elephant species. Therefore, the mathematical models' underestimation of genetic diversity propagation will be greatest for anatomically modern humans. The bottom line is this: nothing in genetics research indicates that the present human population could not have descended from the eight people on board Noah's ark, given the flood's timing as previous to 45,000 years ago (see chapter 11).

While this chapter shows how Noah's flood can plausibly account for the abundance and diversity of life above ground, additional questions remain about how it aligns with what we know of life underground: Earth's biodeposits. How does the quantity and dating of these biodeposits help determine which interpretation of the biblical flood account is correct? This question is the focus of the next chapter.

Chapter 17

The Source of Earth's Biodeposits

Global flood proponents point to Noah's deluge not only as the sculptor of virtually all Earth's geologic features, but also as the preserver of nearly all Earth's astounding wealth of biodeposits.[1] These riches include natural gas, oil, gypsum, topsoil, charcoal, coal, limestone, phosphates, and marble of biological origin. From a global flood perspective, these biodeposits were formed and laid down *not* over our planet's few-billion-year history but, rather, during the 375-day duration of Noah's flood. If the Genesis account were to be understood as a global event, then Earth's biodeposits could be added to the list of fantastic elements.

Biodeposit Quantities

One of the greatest challenges confronting global models of Noah's flood is the sheer quantity of Earth's biodeposits. According to global models, the quantity of Earth's biodeposits would have been determined by the quantity of plant and animal life present on Earth's surface when Noah's flood struck, destroyed, and buried that life.

Common-use fossil fuels make up only a tiny fraction of the biodeposit total. Coal, oil, and natural gas account for less than 0.015%. However, even this tiny percentage poses an insurmountable problem for global flood models. Earth's total known resources of coal, oil, and natural gas prior to the industrial revolution amounted to approximately 9–13 trillion tons (equivalent to 58.8–81.8 trillion barrels of oil).[2]

Non-common-use fossil fuel resources, such as kerogen and clathrates, are far more abundant. Kerogen includes the remains of diatoms, spores, plankton, bacteria, and pollens embedded in sedimentary rocks. Clathrates are a crystalline combination of natural gas and water, formed under below-freezing temperatures or high-pressure conditions, found under oceans and in polar permafrost. Known resources of kerogen measure 900–90,000 trillion tons, and clathrates, 20–7,170 trillion tons.[3]

Sedimentary rock that comprises 7.9% of the total volume of Earth's crust[4] includes a considerable quantity of limestone and marble, at least 6% and more likely about 20%.[5] In other words, limestone and marble most likely make up 1.6% of Earth's crust. Limestone is composed of calcium carbonate crystals from the skeletal fragments of marine organisms such as coral, mollusks, ooids, peloids, intraclasts, extraclasts, and foraminifera, as well as siliceous fragments from diatoms, sponge spicules, and radiolarians.

Earth's continental crust has an average thickness of 35 kilometers.[6] Oceanic crust has an average thickness of 6.5 kilometers.[7] The density of limestone, depending on its porosity, ranges from 1.55 to 2.75 grams/cubic centimeter. The average is about 2.4 grams/cubic centimeter. Organic matter residue in limestone adds up to at least 101,000 trillion tons but, more likely, 337,000 trillion tons.[8]

The total known inventory for just these two biodeposit resources (fossil fuels and biogenic limestone) equals 102,000–434,000 trillion tons. (See the list in table 17.1.)

How do the numbers in table 17.1 compare with the amount of solar energy that Earth's photosynthetic plants, algae, and microbes can maximally capture and convert into biomass? That total adds up to 3×10^{21} Joules, or 500 billion barrels (28 billion tons) of oil equivalent,[9] a quantity that falls short of common-use and non-common-use fossil fuel deposits before the industrial revolution by a factor of 33,200–3,470,000 times. It falls short of Earth's biodeposits of fossil fuels and limestone by a factor of 4,101,000–16,975,000 times.

These shortfalls take into account only the fossil fuel and limestone biodeposits that existed when the industrial revolution began. They do not account for the deposits that have disappeared as a result of degradation, decay, and bacterial consumption. In reality, bacteria continuously consume huge quantities of kerogen.[10] Today, the emission of carbon dioxide to the atmosphere from the bacterial consumption of kerogen nearly equals that from volcanism.[11]

The total of Earth's current living biomass is 625 billion tons.[12] Biological material on and in Earth's crust, just from fossil fuels and biogenic limestone, exceeds Earth's current living biomass by 185,000–766,000 times. This comparison presumes that Earth's living biomass is converted into biodeposits with 100% efficiency when it dies, but it does not. The laws of physics, including the

Table 17.1 Inventory of Earth's Fossil Fuel and Limestone Biodeposits

biodeposit source	quantity (in trillions of tons)
coal, oil, and natural gas	9–13
clathrates	20–7,170
kerogen	900–90,000
organics in limestone	101,000–337,000
total fossil fuel deposits	**930–97,200**
fossil fuels + biogenic limestone	**102,000–434,000**

second law of thermodynamics, or entropy, guarantee that biodeposits are laid down with considerably less than 1% efficiency. As both the Bible and scientific measurements affirm, these laws have remained constant throughout cosmic history (see chapter 18).

Biodeposit Challenge

Even if all Earth's living biomass were continuously converted, upon death, into biodeposits with 100% efficiency, global flood models would fall short in explaining the existence of even a tiny fraction of a percent of Earth's biodeposits. Still more problematic, these models claim that virtually all Earth's biodeposits were laid down in less than a year, during the five months between the beginning of Noah's flood and the ark's coming to rest on the hills/mountains of Ararat.

Recognizing the challenge biodeposits represent,[13] advocates of a global flood offer this three-part response: First, they claim that only a small fraction of Earth's limestone, marble, gypsum, phosphates, clathrates, and kerogen are biological in origin. Second, they suggest that before Noah's flood, one single continent comprised Earth's total land area, and that the one continent was positioned near the equator, where its climate resembled that of today's Amazon rainforest. The climate, in that case, may have been augmented, they say, by some kind of Earth-encompassing canopy, which helped produced a biomass as much as a hundred times greater than what exists on Earth today. Third, they assert that the decaying biological material accumulating on Earth's surface from the time between Adam and the time of Noah (1,656 years, according to their models) was transformed by the flood with 100% efficiency, turning all this material into biodeposits.

No part of this response withstands scientific scrutiny. Biologically derived material exhibits lower ratios of carbon-13 to carbon-12, nitrogen-15 to nitrogen-14, and sulfur-34 to sulfur-32. Geophysicists find pervasive evidence for these lower isotope ratios in the vast majority of Earth's limestone, marble, phosphate, gypsum, charcoal, clathrate, and kerogen deposits.[14] For limestone, in addition to the isotope ratio evidence, paleontologists have found that the skeletal fragments of marine organisms comprise 80–90% of all limestone deposits.[15]

As noted previously, clathrates (methane trapped within a crystal structure of water to form a solid, similar to ice) predominantly occur near continental margins at water depths of 350 to 3,000 meters. At depths ranging from 350 to 3,500 meters, bountiful organic material is deposited into the sediments, and temperature and pressure conditions permit methane arising from organic material to be converted into methane hydrates. If the methane in the hydrates were not mainly biogenic, the clathrate deposits would be much more broadly distributed. As it is, none are found at depths below 5,000 meters.

The second part of the claim, suggesting that Earth's entire terrestrial surface had been a pervasive Amazon-like rainforest with widespread biodeposits evenly distributed throughout all Earth's landmasses fails to match reality. Coal and oil, for example, exist within a few locales on each of the continental landmasses. More than 90% of Earth's continental landmass area is devoid of oil, coal, or natural gas deposits. Similarly, limestone and marble deposits are unevenly distributed. They are found, as previously mentioned, in less than 20% of sedimentary rock formations, and sedimentary rock comprises just 7.9% of the total volume of Earth's crust.[16]

This second part of the response faces yet another problem (besides the implausibility of the canopy hypothesis addressed in chapter 13). Consider the huge quantity of ice on Antarctica and Greenland. If the climate of Antarctica and Greenland had been tropical prior to Noah's flood, then the global flood models allow less than 4,500 years for the formation and accumulation of all Antarctica's and Greenland's ice.[17] Given that this quantity of ice is known to have been relatively stable for the past few thousand years, the global flood models would allow only a few hundred years for the formation of this ice.[18]

The Antarctic ice today measures 4,897 meters (15,650 feet) deep, in its thickest region.[19] The annual snowfall on the continent averages just 165 millimeters (6.5 inches),[20] and in the thickest regions, less than 30 millimeters (1.18 inches).[21] Therefore, even if one assumes that none of the snow and ice in Antarctica were to sublimate (transition from a solid state into a gaseous state without passing through the liquid state) and none were to be compressed by the weight of ice lying above, the thickest part of the present Antarctic ice sheet measures more than 300 times thicker than what any global flood model permits.

The third part of the claim assumes that under benign conditions, with no interruption from natural disasters, dead biological material accumulates without any loss of mass. This, too, is a mistaken assumption. When organisms die, a substantial fraction of their biomass degrades into carbon dioxide, sulfur oxides, nitrous oxides, water vapor, and methane, which dissipate into the atmosphere or into interplanetary space. A much smaller fraction degrades into metal oxides. The dissipation and degradation typically occur rapidly, on timescales of months or years, not millennia.

Meanwhile, detritivores (species that feed on dead biological matter) ensure that most of Earth's dead organic material never becomes part of Earth's biodeposits. Microbes and fungi convert what remains from the detritivores, such as bone and fur, into raw soil nutrients such as carbon, nitrogen, calcium, and magnesium compounds. In other words, even under benign environmental conditions only a tiny percentage of dead biological material accumulates and is preserved as biodeposits.

Importance of Abundant Biodeposits

From a Christian perspective, Earth's wealth of biodeposits shows the design and foresight of a caring Creator. The enormous quantity of biodeposit material residing in Earth's crust provides essential resources for the launch and maintenance of global, high technology civilization. This advanced civilization, maintained for at least a few centuries, makes possible our communication about Christ's offer of salvation to all the world's people groups within a relatively short time.

Today, humanity possesses the technology to lighten dependence on Earth's biodeposits, but this technology could never have been developed without the awesome abundance and diversity of Earth's biodeposits that fueled the industrial and scientific revolutions. The energy resources, building materials, nutrients, and fertilizers that gave rise to the launch and maintenance of a global, large-population civilization were all made possible as solar energy over a few billion years converted dead organisms into usable resources.

Noah's flood, even if it were globally extensive and as maximally energetic as the laws of physics permit, can't come remotely close to explaining the known quantity, extent, and diversity of Earth's biodeposits. For this reason, every global flood model requires radical alteration of the laws and constants of physics—and, on a smaller scale, of Earth's water cycle. The next chapter examines what both the Bible and scientific measurements say about the plausibility of these alterations.

Chapter 18

Rainbows, Physics, and Noah's Flood

When Christian scholars of theology and science reject the historicity of Noah's flood, most do so because they believe it collides with the facts of nature. However, as the preceding chapters attempt to show, nature's realm is not where apparent collisions with the biblical text occur—unless one insists that the flood was global in extent and/or occurred more recently than 45,000 years ago. The clashes occur only within the realm of biblical interpretation.

One of the fundamental principles of biblical interpretation is based on the law of noncontradiction: contradictory propositions cannot both be true in the same sense and at the same time. Therefore, belief that the Bible is wholly inerrant means that no Scripture passage can or will contradict the meaning of any other Scripture passage. An accurate interpretation of Noah's flood described in Genesis 6–11 will be consistent with *everything* the Bible says concerning Noah's flood. Rainbows and physics provide helpful illustrations of how this interpretive rule applies.

The Bible on Rain and Rainbows

Both the Bible and the scientific record have much to say about Earth's water cycle and the physical laws governing the cosmos. A widespread belief among both Christians and non-Christians is that Genesis teaches that no rainbows occurred on Earth until after Noah's flood.[1] This view arises from a misunderstanding of what Genesis 2 adds to the creation account in Genesis 1.

Genesis 2:5 points to an early moment in creation history, a time when neither "shrubs," "plants," "rain," nor "humans" existed. Genesis 2:6 refers to a later time, when "streams came up from the earth and watered the whole surface of the ground," a reference to the Garden of Eden or perhaps a larger region—a time *after* the Creator's establishment of Earth's water cycle (Genesis 1:6–7).

Regardless of locale, the contrast between conditions before and after the operation of the water cycle is unrelated to the conditions before and after the flood. Consider the biblical Hebrew word for rain (*matar*) in Genesis 2:5. It refers to any form of liquid precipitation.[2] The Hebrew word for "streams" (*'ed*) in Genesis 2:6 can mean "mist," "vapor," "stream," or even "flood."[3] The word *'ed*, therefore, can refer to small liquid drops of water in the atmosphere.

Rainbows, of course, are not dependent on the size of liquid water drops. Even the tiniest drops of water in the sky can produce a bright rainbow.

According to Job 36:27–30, 37:13, 38:25–30, Psalm 104:3–6, 147:8, 148:4–8, Proverbs 8:28, Isaiah 55:10, Zechariah 10:1, and Matthew 5:45, clouds and rain have persisted throughout the history of Earth's plant and animal life. In fact, clouds and rain have been present from the second creation day onward. Each of these biblical reflections on the Genesis 1 creation events implies that light—when passing through drops of water in the expanse or sky (Genesis 1:7–8)—produced rainbows, both before and after Noah's flood.

Genesis 9:8–17 refers specifically to God's selection of a sign, or symbol, of his "covenant" with Noah: never again would he destroy all people and all their *nephesh* with a flood. The Scriptures describe five or more covenants (binding commitments) between God and humanity in which God chose something familiar, something previously existing, as a "seal" or symbolic reminder of his promise. The perfect lamb of the Passover promise, and the bread and wine of the new covenant, serve as examples. They affirm the covenant fulfilled in Jesus's bodily death and resurrection. Something as familiar and common as water symbolizes our baptism by and in God's Spirit. Thus, the rainbow fits this pattern—God's choice of something both relevant and familiar to signify his enduring promise.

When God told Noah he would "send rain on the earth" (Genesis 7:4), Noah needed no explanation about the essence of rain. Likewise, Noah seemed to fully understand what God meant by clouds and rainbows (Genesis (9:13–14).

Nature's Record on Rain and Rainbows

Abundant physical evidence shows that rain was falling on Earth long before the human era, even before the existence of plants and animals.[4] Certain sedimentary deposits preserve the splash patterns of ancient raindrops, large and small, just as clearly as other sedimentary deposits show us the pattern of ancient waves lapping the shores of oceans and lakes (see figure 18.1).

Independent methods used to date these splash patterns reveal that

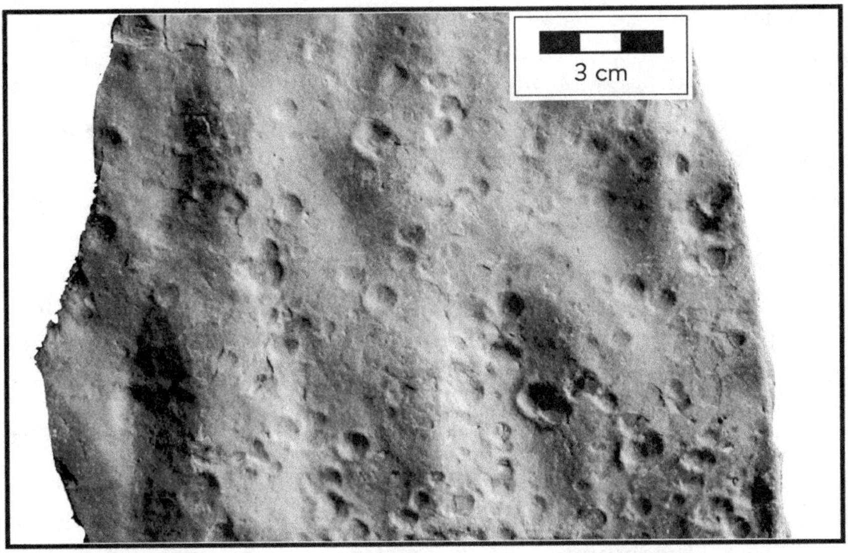

Figure 18.1: Fossil Raindrop Impressions on Mississippian Sandstone
The Mississippian Subperiod is an earlier part of the Carboniferous Period. It lasted from 358.9 to 323.2 million years ago. *Credit: M. C. Rygel, Blue Beach Fossil Museum, Nova Scotia*

rain began to fall prior to the existence of humans and birds and mammals. Fossilized raindrop imprints show that rain was falling in an era when Earth's air pressure was less than half its present level,[5] a level so low as to be severely restrictive to any creatures' habitat range and capacity for activity.

Both the Bible and the geological record affirm that moisture has been falling from the sky throughout the history of plants, animals, and humans. Thus, we can be sure that since the dawn of humanity people have witnessed rainbows.

The Bible on the Laws of Physics

As discussed in chapter 13, global flood proponents are forced to invoke a drastic alteration in the laws of physics at the time of Noah's flood. Both the Bible and the record of nature rule out this alteration.

In Jeremiah 33 we read of God's rebuke to the Jewish nation, voiced through the prophet, for doubting the certainty of his word. God had declared his plan to restore Jerusalem and the nation by raising up "a righteous Branch of David" who would "execute justice and righteousness on the earth."[6] Jeremiah 33:25 compares the certainty of God's promise with "the fixed laws of heaven and earth." As certainly as the laws governing the universe and Earth have operated everywhere without change since they were set in place—laws for which we now have names, such as thermodynamics, gravity, electromagnetism, radioactivity, strong and weak nuclear forces, and more—so God's promises are immutable, guaranteed, and wholly trustworthy.

Romans 8:18–23 reminds us that the universe, since its creation, has been subjected to the constant law of decay. Romans 8:20 says the creation has been subjected to "frustration" or "futility." The next two verses declare that all of creation was and remains in a state of "slavery to decay" or "bondage to corruption."

Ecclesiastes 1 and 3 and Revelation 21 also support this assertion that the whole universe has been subjected to and continues to experience relentless decay. Ecclesiastes 1:1–11 expresses Solomon's lament over the ongoing "sameness" of things. The second and third chapters of Genesis imply that work and pain, though substantially more difficult under sin's curse, were part of the creation even before sin occurred. Heat consistently flows in the same direction—from hot to cold. "Bondage to decay" aptly portrays the second law of thermodynamics as the physical universe becomes increasingly disordered. While this law may seem a negative aspect of God's creation, it isn't. The progression to disorder (entropy) is essential for physical life. This law makes digestion of food possible. It makes physical work possible.

Romans 8:18–23 assures us that the laws of physics will remain in effect until God's work of redemption has been completed. God's stated plan is to bring a vast host of humanity from every "tribe" (or ethnicity) on Earth into his own family, an eternal fellowship under the authority of Jesus Christ. This work will be done only when the full number of humans God intends to redeem has repented of sin and, by grace, through his gift of faith, received adoption as his own.

Among written works or concepts on which the world's religions rest, the Bible stands alone in affirming that the laws God set in place to govern the universe and Earth are constant. The immutability of these laws is what makes scientific investigation of the natural realm possible. This investigation can and will reveal truth because of that constancy. Thus, the Bible assures us that thorough, honest scientific study of Noah's flood can be trusted.

Nature's Record on the Laws of Physics

Many independent physics laboratory experiments directly verify the operational characteristics and values of the laws of physics. Physicists have developed instruments capable of measuring the constants associated with the laws of physics to a precision that, in some cases, provides accuracy to 18 places of the decimal.

The discipline of astronomy is unique among the sciences in that it allows for direct access to the past. This access is a consequence of the known and measured velocity of light, neutrinos, and gravity waves, a velocity proved constant to within 15 places of the decimal.[7] Due to this constancy, astronomers can directly measure the characteristics and operation of the laws of physics at any epoch during the history of the universe simply by choosing to observe astronomical objects at varying distances from Earth. The greater the distance of an astronomical object, the farther back in time we are seeing.

Constant Velocity of Light

For astronomical observations to be a reliable indicator of the past, the velocity of light/gravity/neutrinos must be immutable (an unchanged fixed value) and precise. That reliability also requires a precise velocity measurement.

In 1973, the National Bureau of Standards published a measurement of the speed of light that was accurate to within 3 parts in a billion (a precision of 0.0000003%). The determined value—based on the fundamental standard of time (cesium atomic clock) and of length (krypton-86 radiation from an electric discharge lamp)—was shown to be 288,792,456.2 ± 1.1 meters/second.[8] Since 1983, the International System of Units has defined the velocity of light as exactly 299,792,458 meters/second, using this value to precisely establish the length of a meter.

Astronomers have developed three independent ways to establish that the speed of light has remained unchanged throughout the history of the universe. One such way is by observing both the light and heat emitted by the Sun. Even a minuscule change in the speed of light would alter the Sun's luminosity (brightness), as well as the relative abundance of elements produced in the Sun's nuclear furnace. The *c* in Einstein's famous equation for special relativity, $E = mc^2$, stands for the speed of light. A change in the value of *c* would yield geometrically greater changes in the *m* (matter) or *E* (energy) or both. Thus, even a mere 2% change in the Sun's luminosity would be enough to exterminate all plant and animal life on Earth.[9] Altering the quantity of greenhouse gases in Earth's atmosphere to compensate for a change in the Sun's luminosity would allow for a greater change, but not by much. For example, even if Earth's atmospheric carbon dioxide level were reduced to the minimum level required for photosynthesis, all plant and animal life on Earth would die because the Sun's luminosity would be increased by 4%.

Scientists have also demonstrated the constancy of light's speed using vanadium and copper. The production of vanadium- and copper-based proteins depends critically on the value of the relativistic time dilation factor ($\{1 - v^2/c^2\}^{-0.5}$), which depends critically on the value of light's speed. Relativistic time dilation, if altered by the tiniest amount, would prevent vanadium- and copper-based proteins from fulfilling their critical roles in higher animals' nervous system functioning and bone development.[10]

A third way to establish light's constant speed is to measure the wavelength positions of certain spectral lines—those shifted by interactions with electric and magnetic fields—in distant stars and galaxies. When astronomers do so, they determine the value of the fine-structure constant at the moment light was emitted from a star or galaxy. (The fine-structure constant quantifies the strength of the electromagnetic force interactions between elementary charged particles.) The more distant the star or galaxy, the farther back in time astronomers can measure the fine-structure constant value.

The fine-structure constant value depends inversely on the value of light's speed. Thousands of measurements made by astronomers show that the fine-structure constant in stars, galaxies, and quasars up to 12.9 billion light-years away differs by no more than 1 part in 1,000,000 from Earth-based measurements.[11] This finding implies that possible variations in the value of the fine-structure constant amount to less than 7.8×10^{-17}/year (less than 0.0000000000000078%) throughout the past 12.9 billion years. Laboratory experiments using single-ion optical clocks conducted over a span of a year yielded an even greater constraint, showing that possible variations in the fine-structure constant are less than 3.9×10^{-17}/year.[12]

The fine-structure constant also depends on the square of the electron charge value and the inverse of the value of Planck's constant. Planck's constant relates the energy of a photon to its frequency. A small change in the value of Planck's constant leads to devastating consequences for life. Such a change would greatly disturb the relative abundances of life-essential elements.[13]

Technically, the speed of light could change while the fine-structure constant remains unchanged *if* the electron charge value were to change enough to compensate for the change in light's speed. However, even a tiny change in the electron charge value would result in catastrophic consequences for life. It would force a violation of the second law of thermodynamics sufficient to disrupt photosynthesis in plants and digestion in animals.[14] It would disrupt the stability of the Sun and all other stars, given that stellar luminosity stability critically depends on a delicate balance between gravity and the electron charge value.[15]

Natural Nuclear Reactor Test for Constant Physics

Two billion years ago, when uranium-235 was more abundant, a natural nuclear fission reactor functioned in Oklo, Gabon (western Africa). Calculations performed on fresh cores in this reactor with different uranium isotope contents, combined with measurements of the samarium-149 cross section and the ratios of samarium-147 to samarium-149, established that changes in the fine-structure constant and the strange quark mass over the past 2 billion years were less than 1.0×10^{-17}/year and 1.0×10^{-18}/year, respectively.[16]

Cold Molecule and Atomic Clock Tests

Experiments on ultracold molecules (colder than 140 microKelvin or 0.0014°C above absolute zero) conducted over a 4-year period showed that variation in the electron-to-proton mass ratio must be less than 1.3×10^{-14}/year.[17] Experiments comparing Ytterbium$^+$ and Cesium atomic clocks revealed that the electron-to-proton mass ratio has varied by less than 2.1×10^{-16}/year.[18] A team of ten British physicists using single-ion atomic optical clocks determined that variation of the electron-to-proton magnetic moment ratio must be less than 1.3×10^{-16}/year.[19]

Verification Made Possible by Apollo Missions

On July 21, 1969, millions watched live on television as Apollo 11 astronauts Neil Armstrong and Buzz Aldrin set up a laser reflector on the lunar surface, an instrument to be used in testing the constancy of gravity (figure 18.2). In 1971, the Apollo 14 crew placed a similar laser reflector on a different lunar site. Later that year, the Apollo 15 crew set up yet another laser reflector, this one three times the size of the one installed by Armstrong and Aldrin. Figure 18.3 shows the locations of the three laser reflectors.

Over the past 51 years, six observatories located across the United States and Europe have been regularly bouncing laser reflections off the three Apollo lunar laser ranging instruments to determine the gravitational force constant. These experiments have shown that the gravitational force constant varies by less than 1.5×10^{-14}/year.[20] This finding also constrains possible variations in the electron-to-proton mass ratio to less than 4.5×10^{-14}/year.[21]

Figure 18.2: Apollo 11 Lunar Laser Ranging Instrument
Credit: NASA

Observational Verifications of Immutable Physical Laws

As noted previously, astronomical observations establish that the value of the fine-structure constant has changed by less than 7.8 x 10^{-17}/year throughout the past 12.9 billion years. Astronomers' observations of spectral lines in stars, galaxies, and quasars show that variations in the electron-to-proton mass ratio are less than 1.36 x 10^{-17}/year over the past 7.34 billion years[22] and less than 1.6 x 10^{-17}/year throughout the past 12.9 billion years.[23]

Similarly, astronomical observations determine that the gravitational force constant has not varied by more than 7.9 x 10^{-12}/year over the past 11.0 billion years.[24] Pulsar timing observations conducted over a 26-year period have established that any changes in the gravitational force constant must be less than 1.0 x 10^{-12}/year.[25] Helioseismology observations conducted over a 24-year period demonstrate that the variation in the gravitational force constant must be less than 5.2 x 10^{-14}/year.[26]

These exacting measurements may be difficult for nonscientists to

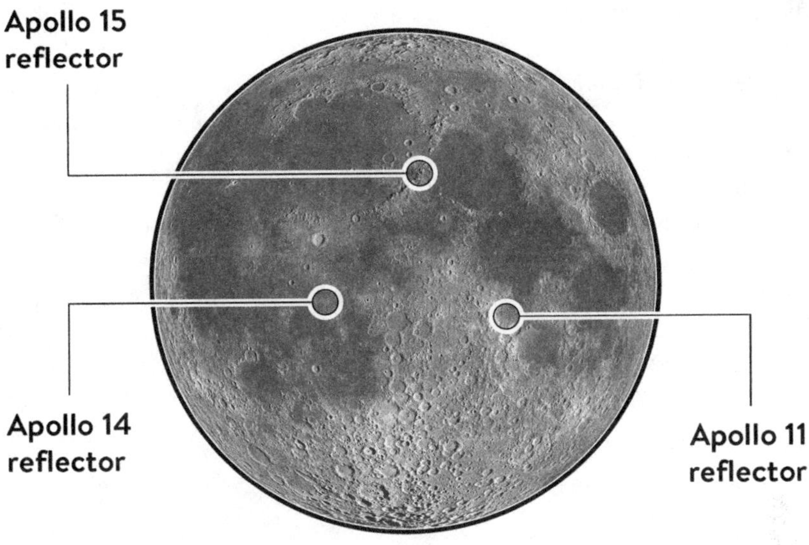

Figure 18.3: Lunar Laser Ranging Reflector Sites
Image credit: NASA; diagram credit: Hugh Ross

comprehend or appreciate. One way to think about them is to consider that scientific advances have allowed for increasingly precise measurements at greater lookback times. That increasing precision in our view of the past, in turn, provides increasing affirmation that the Bible teaches truth in all matters of science it addresses. It also affirms that the Bible has accurately predicted many scientific discoveries.

Several papers in the scientific literature speculate that minor changes in a few of the constants of physics might have occurred when the universe was less than 1 second old, but astronomers lack tools for testing physics in that ultrabrief era. Nevertheless, these speculations are irrelevant to interpretations of Noah's flood, which took place during Earth's history. The scientific experiments and observations cited previously apply to the entire history of Earth.

Implications for Noah's Flood

A literal, consistent interpretation of biblical texts relevant to the laws of physics, including light's interaction with water, affirms beyond reasonable doubt that the physical laws governing Earth and Earth's life have persisted throughout the entire history of life's existence. This affirmation is firmly sustained by scientists' experiments and observations. On this basis, one can reasonably conclude that all interpretations of Noah's flood that depend on localized or cosmic-scale changes in the laws of physics before, during, or after Noah's flood are invalid. This conclusion applies to all global flood models.

The constancy of the physical laws provides yet another strand of evidence for a regional, yet universal flood. Having established a biblical, historical, and scientific basis for the veracity of Noah's flood, we now consider a vital question for all of us: What lessons can we learn from Noah's flood? The next chapter addresses this question.

Chapter 19

Purposes and Benefits of Noah's Flood

To accept the account of Noah's flood as historical narrative may seem one of the greatest challenges—if not *the* greatest—to belief and trust in the loving, merciful, all-knowing, all-powerful personal God the Bible reveals. Some would argue that a good and gracious God could have found some other way to deal with humanity's collapse into pervasive evil than to bring utter devastation on all but one family—an imperfect one, at that.

Of course, God could have created humanity without free will—or with *limited* free will and, in that case, a lesser inclination to make conscious choices. Consider, however, what our relationships would be like in such a world. While selfishness might be expressed to a far lesser degree, so would love. God wanted strong love, not weak love relationships with and among his "image-bearers." Strong free will makes strong love possible.

Some might say God could have intervened at an earlier stage in humanity's descent toward depravity, but the God of Scripture is perfect in patience. He gave people, as he still does, every possible chance to turn things around. Consider the people's extended exposure to Noah's preaching—and to the growing witness of the enormous ark itself. God's patience gives everyone as much time as possible to repent. As we read in 2 Peter 3:9, "The Lord is not slow . . . as some understand slowness. Instead he is patient with you, not wanting anyone to perish, but everyone to come to repentance." Nevertheless, he knows when that final moment has been reached. The fact that we lack full understanding of God's timing and plans and reasons for action or inaction does not mean God lacks good reasons.

Vital Lessons from Noah's Flood

We humans tend to minimize the gravity of our sin and the weight of its damage to our relationships—with God, ourselves, other people, other creatures, and our planet. The human condition at the time of Noah's flood shows us, vividly, the consequences of failure to recognize and restrain sin. Clearly, we need God's help. God's command to Noah and his family when they disembarked from the ark specifically addressed the need to deal quickly and justly with the sin of murder. In Genesis 9:5–6, we see God's emphatic statement that murder must no longer be tolerated. Human authority structures must do whatever is necessary to restrain murder.

In the command to restrain murder, God reminds Noah, his family, and their descendants that every human being bears his likeness. This reminder emphasizes that in addition to restraining murder, human authorities must do everything within their power to ensure that each person is treated with appropriate dignity and respect. Everyone has inherent value as a person made by God and for God. Therefore, abusive treatment of anyone—child or adult, poor or rich, foreigner, minority person, and someone with mental or physical differences in ability—must not be tolerated.

God's shortening of the human lifespan underscores our need for divine help in restraining sin. As explained in chapter 15, a lifespan in the hundreds of years, which many people seem to think would be wonderful, was demonstrated to cause more harm than good. In the context of human free will and our propensity to sin, lifespans of such length facilitate the spread of evil. A maximum lifespan of about 120 years provides sufficient time for individuals to decide who or what they will devote themselves to and, thus, with whom they choose to spend eternity. Jesus showed us the reality of life beyond the limits of our earthly existence.

Genesis 6:4 notes that "the Nephilim were on the earth in those days [the days before the flood]—and also afterward—when the sons of God went to the daughters of humans and had children by them." According to ten different Bible passages,[1] these men of towering stature and great strength (also known as the sons of Rapha, the Rephaim, sons of Anak, Anakites, Anakim, giants, or heroes of old) were thoroughly evil, a threat to humanity's survival.[2]

The Hebrew word *Nephilim* means "fallen ones." For the purposes of this book, it's sufficient to note that because of their exceptional wickedness and prowess, they were super-spreaders of reprehensible, reprobate behavior among human societies. Genesis chapters 6–9 show us that evil tends to metastasize, like a deadly cancer. Noah's flood rid the earth of these beings. Later, because some of these "sons" again went to the daughters of humans to produce another generation of Nephilim, God raised up Moses, Joshua, then David and David's "mighty men" to wipe them out for good.

God used the flood to prevent the self-extermination of human beings. Our extinction appeared imminent, and the human species was not the only life-form at risk. The *nephesh* (soulish) creatures in close contact with depraved humans also suffered irreparable damage. This damage, like the damage to humans, was in grave danger of spreading even further. In this context, Noah's flood can be seen as a merciful rescue operation.

Following the story of Noah's flood and the account of humanity's long-delayed spread from a single region of the planet (Genesis 6–11) comes the story of Abraham and of God's covenant with us through him (Genesis 11–15). As noted in chapters 5–6, Christian scholars throughout all generations of the church have recognized and taught the symbolic significance of the Genesis account of Noah's flood. It foreshadows the gospel message, God's gift of redemption by grace through faith.

Nearly all the major Old Testament narratives, while historically true, also provide a preview, an unmistakable portrayal, of the future work of Jesus Christ. They show us how people who chose to put their trust in God were rescued from bondage to sin and death and granted eternal fellowship with the Father, Son, and Holy Spirit. Of all the Old Testament stories, this account of Noah, the ark, and the rescue of Noah, his family, and their animals through the flood provides one of the clearest and most dramatic depictions of the gospel message.

Earthly Benefits of Noah's Flood

Just as Scripture reveals at least some of the spiritual purposes and benefits of Noah's flood, God's book of nature also reveals important benefits. Science has allowed us to discern at least a few. Perhaps more will be recognized in future years.

The runoff of rain, melting snow, and melting ice from mountains, hills, and plateaus onto the plains surrounding the flood region deposited vast quantities of soil nutrients. This fertilization event was likely accompanied by another boost. Whenever thick ice sheets melt, the crust below them rebounds from being pressed down by their heavy weight. These rebounds, in turn, give rise to widespread volcanic activity, producing essential soil nutrients.[3]

The fertilization of agricultural plains nearly all over the world would help explain, at least in part, how Noah's descendants were able to relaunch civilization as quickly and as aggressively as they did. This relaunch set the stage for population growth, as well as for the spread of humanity to previously uninhabited parts of the globe. Finally, humanity could make progress in fulfilling God's assignment (Genesis 1:26–30) to multiply, fill the earth, and wisely manage its resources for the benefit of all life.

The timing of Noah's flood appears providential in another way. To spread out over the whole planet, people needed the land bridges that once (and for a limited time, only) connected Earth's continents to each other and to large islands. As facilitators of human migration and settlement of Earth's continents, these bridges had to match human requirements with respect to climate, terrain, flora, and fauna. As described in chapter 11, Noah's flood and the events soon afterward, including God's supernatural "nudge" (at Babel), occurred at the last moment, while those bridges were still intact. Recent research confirms that just-right land bridges allowed for human migration into southern Europe, northern Europe, all of eastern Asia, Australia, New Guinea, and Tasmania—rapidly and nearly simultaneously.

It seems possible, even likely, that Earth benefited in other ways from the timing and location of Noah's flood. However, because the flood occurred during an ice age—likely the most recent one in the cycle (see chapter 11)—the geological and archaeological evidence from it would have been erased by rapid and widespread melting and reforming of ice sheets as well as by the accompanying erosion and volcanic activity. One can only hope that ongoing research will tell us more.

The Flood's Most Important Lesson

The story of Noah's flood, as chapter 3 documents, is known—in one form or another—among virtually all cultures and people groups of the world. Even in its distorted forms, it raises questions about the human condition and more. Its meaning and significance are available for all to ponder.

Water plays a dual role in the Bible. It brings God's judgment on his enemies but it also brings salvation to his own. From beginning to end, God's redemption story portrays the Spirit of God as life-giving water. We see this water in the Revelation of Jesus, "as clear as crystal, flowing from the throne of God and of the Lamb, down the middle of the great street of the city," the holy and eternal city of God (Revelation 22:1–2).

Those of us who have access to the Scriptures have a responsibility to carefully consider whether the biblical account of Noah's flood is historically defensible or not. If we can show it to be true, then we remove a major challenge to biblical inerrancy and a stumbling block to faith. The door will be opened for Bible readers to affirm that everything written in the Bible's 66 books on all the topics they address has been supernaturally inspired by God to convey truth.[4]

Confidence that the Bible is, indeed, the true and trustworthy Word of God provides powerful motivation to consider what its stories, including the account of Noah's flood, reveal about the heart of God. Just as God reached out with precise timing and a perfect plan to rescue Noah and his family from a deadly flood, so God reaches out with precise timing and a perfect plan to rescue from spiritual death all who put their trust in him through Jesus Christ—God's Son, our Creator and Redeemer.

Appendix

Dating Early Human History

One of the most frequent inquiries about how to interpret the early chapters of Genesis concerns the timing of the events recorded there. People especially want answers to these questions:

1. When did Adam and Eve exist?
2. When did Noah live?
3. What is the date for Noah's flood?
4. When did God scatter people over all the earth?
5. When did the diversification of languages occur?

Determining scientific dates for these events is a challenge, and estimates are imprecise, but not for a lack of effort on the part of researchers. All scientific methods applied to dating events in the early history of anatomically modern humans—as distinct from the hominins that preceded *Homo sapiens sapiens*—involve significant statistical errors and even larger systematic errors.

Kinds of Errors

Statistical errors, also known as random errors, refer to imprecision in making measurements. *Systematic* errors refer to environmental and instrumental factors that could shift all the measurements of an entity up or down in value by the same amount.

In some cases, the environmental and instrumental factors are known. In these instances, scientists can determine the range of possible alterations to the value of the measurements. In other cases, scientists are aware that environmental and instrumental factors are involved, but they're unable to determine the range of possible alterations to the measured values. In the remaining cases, scientists can identify only a few or none of the possible environmental and instrumental factors impacting measurements.

Simple Sciences Contrasted with Life Sciences

In the simple sciences, such as astronomy, physics, physical chemistry, and geophysics, environmental and instrumental (systematic) factors typically are well understood, and scientists are able to determine a range of values for possible alteration of their measured values. Thus, when physical scientists publish a research paper, their results include an average of their measurements, plus or minus the probable statistical error and plus or minus the probable systematic error. Unlike the statistical error, in which the plus value equals the minus, for the systematic errors, the value of plus and minus may differ.

Unless otherwise stated, the probable error, statistical and systematic, in the simple science findings represents a 67% certainty that the actual value falls within the stated error range. Further, scientists working in the simple sciences often present both the statistical and systematic errors in their results even where the stated probable error range represents a 95 or 99% certainty that the true value falls within the stated error range.

A rule of thumb in the simple sciences is that no published result is to be trusted unless the authors take these two steps: identify the full range of possible systematic effects, and demonstrate how they determined the range of possible alterations caused by the systematic effects. Even then, caution is in order. In the astronomy and physics journals, readers frequently see examples of papers pointing out one or more systematic effects overlooked in previously published papers.

In the life sciences, the subject matter is typically so complex that researchers are unable to identify the full range of possible systematic effects, let alone the range of alterations such effects might cause. Therefore, life scientists typically publish *only* their known statistical errors. Those who read the life science research literature often place more confidence in the published results than the results warrant. In turn, the lay public tends to trust published life science results more than the data and/or conclusions drawn from the data warrant.

Carbon-14 Dating Limitations

With respect to human history, the most reliable method for dating events is the carbon-14 method. It's the *only* radiometric tool that's useful for dating human remains and human artifacts. Carbon-14 dating measures how much time has passed since a living organism has stopped taking in carbon molecules from the atmosphere. In other words, it measures how long an organism or tissue from it has been dead.

The half-life of carbon-14 is 5,715 ± 30 years.[1] Apart from the rare case when researchers are dealing with a large pure sample, dates based on radioactive decay are reliable within the range of one-eighth to eight times the carbon-14 half-life of 5,715 years. In other words, carbon-14 measurements are most accurate when used to date samples 700–45,000 years old. Dating of samples outside that likely age range cease to be reliable. Thus, any sample older than 45,000 years cannot be reliably dated using carbon-14. Even within the appropriate range, however, carbon-14 may not be free of error from systematic effects, as the following three examples show.

Cosmic Ray Effect

Carbon-14 in the atmosphere is produced when cosmic rays strike nitrogen-14. One systematic effect in carbon-14 dating is that the quantity of cosmic rays reaching Earth's atmosphere can change over time. Cosmic rays predominantly come from the remnants of supernova eruptions and from supermassive black holes. Because of the uneven distribution of supernova remnants and supermassive black holes on the celestial sphere, the rate (abundance per minute) with which cosmic rays reach Earth's atmosphere varies slightly on a yearly basis. During the past 50,000 years, at least five supernova events have occurred within 250–1,000 light-years from Earth.[2] Those events were close enough to have altered carbon-14 dates, although by no more than 10%.

Fortunately, the alterations resulting from variable cosmic ray exposure often can be accurately determined. The quantity of carbon-14 measured in the annual layers in tree rings, varves, and ice cores allows scientists to accurately establish the varying rates at which cosmic rays strike Earth's atmosphere. Where such measurements are available, scientists refer to the carbon-14 date as a "calibrated carbon-14 date." The latest calibration curve is IntCal20.

Location of Organism

The location of an organism during its lifetime can also lead to systematic effects in carbon-14 dating. If it lived at a high altitude, the air it was taking in would have been exposed to more cosmic rays. If it lived underground or under a dense forest canopy, the air it took in would have been exposed to fewer cosmic rays.

Radioisotope Exposure

All diamonds and zircons contain small amounts of nitrogen-14. Nearly all contain mineral inclusions wherein reside small quantities of radioisotopes such as potassium-40, uranium-235, uranium-238, and thorium-232. The decay of these radioisotopes transforms some of the nitrogen-14 in diamonds and zircons into carbon-14. The presence of nitrogen-14 and radioisotopes explains, for example, why zircons and diamonds—though billions of years old—yield carbon-14 dates of about 58,000 years.

On the positive side, scientists can almost always identify the systematic effects in carbon-14 dating and determine the resultant errors. Thus, where carbon-14 dating is applicable, it remains the method anthropologists prefer as they investigate human history and human artifacts. It's the *only* reliable method at their disposal and is useful *only* for dating human remains and artifacts younger than about 45,000 years old.

Limits on Other Dating Methods

When carbon-14 dating is not an option, researchers must use indirect dating techniques for determining age. These alternative methods, described in the following section, are prone to large systematic effects that in most cases cannot be measured.

Thermal and Optical Luminescence

Heat and light cause certain chemicals in material or tissue to fluoresce. A sample's capacity for fluorescence gives an indication of how long that sample has been buried, or otherwise cut off from heat and light/sunlight. For example, when a crystalline grain of quartz is buried and removed from solar exposure, the radioactive decay of uranium and thorium in surrounding rocks and soil forces electrons in that crystal out of position. Some of these electrons accumulate over time in defects within the crystal. Optical luminescence measures the degree of electron buildup to determine how long ago the crystal was buried.

Thermal and optical luminescence suffers from obvious systematic effects, often large ones. For instance, the intensity of the heat and/or light to which the sample was exposed prior to burial may vary greatly, depending on its preburial environment and the rate at which its burial progressed, whether immediate or prolonged. The various times and ways a sample's burial was interrupted also alters its age measurement.

In thermal and optical luminescence, researchers date one or more mineral crystals in an artifact rather than the remains of an organism. Researchers must assume how the artifact arrived in its current location—whether it was brought there by an ancient human or by some other means. They must also attempt to distinguish between a crystal associated with a relatively recent burial and an artifact from crystals buried earlier. They face the added challenge that an artifact buried in sediment of a relatively young age may have been moved there from a much older sediment by one or more geological processes or by the actions of some other creature—such as the digging activity of a bear.

A famous example of the enormity of possible systematic effects in thermal and optical luminescence comes from the dating of artifacts in the Jinmium Rock Shelter in northern Australia. A thermoluminescence date of aboriginal artifacts was cited as evidence that humans had occupied Australia for at least the past 60,000 years.[3] However, a later carbon-14 analysis showed that the oldest artifacts were only 3,000 years old.[4] In the words of the authors of the latter analysis, the earlier date was off "by more than an order of magnitude."[5]

This reassessment of the Jinmium artifacts does not rule out the presence of humans in Australia earlier than 3,000 years ago. Nevertheless, it does indicate that the thermoluminescence date is extremely inaccurate. At other sites in Australia where both radiocarbon dating and atomic mass spectrometry have been applied, studies have dated human artifacts as far back as about 30,000 years.[6]

Electron Spin Resonance

Another dating technique employed by anthropologists is electron spin resonance—a sophisticated version of the thermal and optical luminescence method. This technique involves measuring the number-density of unpaired electrons in a sample that's been previously exposed to natural radiation. For this method to yield a reliable date, a researcher would have to know the history of that sample's exposure to natural radiation, including the radiation source(s) and the duration of exposure. In most cases, this history is unknown. Thus, dating by electron spin resonance is subject to the same systematic effects that plague thermal and optical luminescence dating.

Uranium-Thorium Dating

The uranium-thorium dating method measures how much time has elapsed since a sample first precipitated (separated) from water. The fact that thorium is water soluble and uranium is not makes this method possible. One of the rare isotopes of uranium, uranium-234, decays into thorium-230 with a half-life of 245,000 years. (Uranium-234 exists on Earth despite the 4.567-billion-year age of Earth because it is an indirect decay product of uranium-238, which has a half life of 4.468 billion years.) Determining the ratio of thorium-230 to uranium-234 in a sample yields the time since its precipitation if, and only if: (1) the researcher knows that the sample comes entirely from a single rapid precipitation event, and (2) the sample hasn't suffered significant disturbance or contamination since it precipitated.

DNA Dating

Two of the most frequently cited tools for dating events in early human history are mitochondrial DNA analysis and Y-chromosomal analysis. Because our mitochondrial DNA comes exclusively from our biological mothers and males' Y chromosome comes exclusively from their fathers, geneticists can determine the time back to the earliest female and the earliest male from whom all humanity has descended. The process involves these three steps:

1. measuring the genetic diversity in the present human population,
2. assuming mutation rates for mitochondrial DNA and Y-chromosomal DNA, and
3. assuming an average time between birth and reproduction.

Some of the possible systematic errors in DNA dating are obvious to researchers, while others seem more subtle. Scientists know that the human mutation rate can and does vary across time and geographical regions. Certain environmental and social factors are known to substantially impact the human mutation rate. A host of other factors have yet to be studied or fully understood.

Astronomical Uncertainties

Even the known environmental and social factors involve enormous uncertainties. For example, mutations can be generated by cosmic rays from past supernova eruption events and/or by past major solar flares. During the last 10,000 years, no supernova eruptions closer than 5,000 light-years from Earth have occurred.[7] However, at least four supernova eruptions occurred between 20,000 and 50,000 years ago just 300–900 light-years distant.[8] Many more occurred between 50,000 and 115,000 years ago.[9] As for solar flares, no major ones have impacted Earth during the past 10,000 years, but several likely occurred between 10,000 and 115,000 years ago. Failure to consider the impact of supernova eruptions and major solar flares means that published DNA dates for human origins and events in early human history may well be seriously overestimated.

Human Reproduction Uncertainties

Likewise, no one can determine with accuracy the time span between human birth and reproduction throughout human history, especially given potential increases and decreases in human longevity. However, estimates that incorporate today's widespread availability of birth control and extended career-launch time before childbearing certainly put that value at the high end.

From a biblical perspective, it's possible that God created Eve with a diversity of mitochondrial DNA in her ova. In that case, mitochondrial DNA time estimates back to the first woman may be significantly briefer than current calculations indicate. Similarly, the shortening of potential human lifespans that occurred after Noah's flood may have altered Y-chromosome DNA.

Dating the Prehuman Hominids and Their Artifacts

For prehuman remains and artifacts, a few more reliable dating methods are available. For example, argon-argon dating and paleomagnetic dating can be employed on samples ranging from 250,000 years ago to several million years ago. These methods, though not without systematic effects, yield much smaller errors than those used in dating early human history. This difference explains why readers can place somewhat greater confidence in the published dates for the prehuman hominids.

Four Cautions on Scientific Dates for Early Human History

First, readers must be alert to "edge-of-the-error-bar" bias. An example of this tendency may be seen when scientists, journalists, and/or internet bloggers claim that humans date back to at least 200,000 years ago based on a published calculation that mitochondrial Eve lived 157,000 ± 40,000 years ago. Citing the date as 200,000 years ago (the upper edge of the error bar) is not the same as presenting the date as 197,000–117,000 years ago (the range).

Second, readers need to realize that many published dates for early human events and artifacts include *only* the statistical, not the systematic, error bar. Often, these probable errors are less than 10% of the claimed date range. When readers see such a small statistical error, they tend to conclude that the claimed age can be trusted as reliable when, in fact, the systematic error could be larger than ± 1,000%.

Third, readers need to recognize that for anatomically modern humans, none of the methods available to scientists (apart from carbon-14 dating) allow them to determine the value of likely systematic errors. The inability to make this determination explains why anthropologists refrain from publishing *both* statistical *and* systematic effects for their age measurements. However, anthropologists will, on occasion, acknowledge (in their peer-reviewed papers) the potential presence of systematic effects, though without attaching a value to them. The caution shown by these authors is appreciated, but readers would still be wise to exercise caution, recognizing that anthropology is so extremely complex that no anthropologist can be expected to identify all likely and/or possible systematic effects, whatever dating method they use.

Fourth, as the examples mentioned show, systematic effects for age measurements relative to anthropology nearly always point in one direction, toward the older date. That is, the actual age is more likely to be substantially lesser than greater. Therefore, wherever large systematic effects are known or anticipated, the reader should consider the stated age as an approximate upper limit.

An appreciation for how statistical and systematic effects impact the dating of objects and events in early human history gives scientists reason for tentativeness and humility. Though scientists strive diligently to solve difficulties, it appears that for now, at least, determination of precise dating with respect to early human history remains elusive.

Tentative Human History Answers

With all these cautions in mind, some of the five questions raised at the beginning of this appendix can be answered more accurately than others. The scattering of people over all the earth falls within the range of carbon-14 dating, though near the limit of its applicability. As explained in chapter 11, a trustworthy date of 45,000±5,000 years ago can be assigned to this event. From a biblical perspective, the diversification of languages occurred at the same time as the scattering of people over all the earth. However, all that can be established scientifically is that diversification of human languages occurred previous to 17,000–13,000 years ago.

Answers to the first three questions all fall outside the range of carbon-14 dating and, therefore, remain much less certain. Clearly, Noah's flood and Noah's birthdate must have predated the scattering of humanity. Genesis 7–8 provides indicators that Noah's flood occurred during an ice age, most probably the most recent ice age. Layers in deep ice cores establish reliable dates for that ice age. It endured from 120,000 to 12,000 years ago. Therefore, Noah's flood likely occurred between 120,000 and 45,000 years ago. A case can be made that Noah's flood occurred during one of the major, rapid melt events of that ice age. More than a dozen such events occurred from 80,000–45,000 years ago. Somewhere within this date range seems a reasonable estimate for the timing of Noah's flood.

The least certain human history date is that for the creation of Adam and Eve.

The best mitochondrial and Y-chromosome DNA dates range from 204,000 to 98,000 years ago. However, as pointed out in chapter 16, these dates fail to take into account several systematic effects that are either unknown or undetermined.

Notes

Chapter 1: A Profound Stumbling Block

1. Bert Thompson, "In Defense of the Genesis Flood," *Apologiaway: Conservative Christian Worldview* (July 26, 2019), accessed August 24, 2023, apologiaway1.wordpress.com/2019/07/26/in-defense-of-the-genesis-flood/.
2. Donald R. Prothero, "A Gentle Journey Through the Truth in Rocks," *eSkeptic*, January 2, 2013, accessed October 17, 2023, skeptic.com/eskeptic/13-01-02/#feature.
3. Hugh Ross, *Navigating Genesis: A Scientist's Journey Through Genesis 1–11* (RTB Press, 2014), 139–182.
4. Cary Funk and Becka A. Alper, Pew Research Center, "Religion and Science," (October 22, 2015), accessed June 5, 2023, pewresearch.org/internet/wp-content/uploads/sites/9/2015/10/PI_2015-10-22_religion-and-science_FINAL.pdf.
5. Ronald L. Numbers, *The Creationists: The Evolution of Scientific Creationism* (Alfred A. Knopf, 1992), 72–139, 187–213.
6. Ken Ham and Bodie Hodge, *A Flood of Evidence: 40 Reasons Noah and the Ark Still Matter* (Master Books, 2016), 57.
7. Ham and Hodge, *A Flood of Evidence*, 58.
8. Augustine of Hippo, *The Literal Meaning of Genesis*, book 1, chap. 19, paragraph 39 in "St. Augustine: The Literal Meaning of Genesis," *Ancient Christian Writers*, nos. 41–42, trans. J. H. Taylor (Newman Press, 1982), 42–43.
9. Hugh Ross, *Rescuing Inerrancy: A Scientific Defense* (RTB Press, 2023), 32–48.
10. William Ackerman et al., "The Chicago Statement on Biblical Inerrancy," *International Council on Biblical Inerrancy* (January 1, 1979), accessed August 24, 2023, library.dts.edu/Pages/TL/Special/ICBI_1.pdf; library.dts.edu/Pages/TL/Special/ICBI_1_typed.pdf.
11. J. I. Packer et al., "The Chicago Statement on Biblical Hermeneutics," *International Council on Biblical Inerrancy* (November 13, 1982), accessed August 24, 2023, library.dts.edu/Pages/TL/Special/ICBI_2.pdf.
12. Ross, *Rescuing Inerrancy*, 67–209.
13. Ross, *Rescuing Inerrancy*, 126, 158–160, 175–17.

Chapter 2: The Basic Story

1. Genesis 6:13–14, 17, 7:20, 24, 8:15–17.
2. R. Laird Harris, Gleason L. Archer Jr., and Bruce K. Waltke, *Theological Wordbook of the Old Testament*, vol. 1 (Moody Press, 1980), 297.

3. Genesis 5:32, 7:6, 11.
4. Hebrews 11:7; 2 Peter 2:5.

Chapter 3: Everywhere a Flood Story

1. Johannes Karl Richard Riem, *Die Sintflut in Sage und Wissenschaft* (Agentur des Rauhen Haufes, 1925), scribd.com/document/603003993/Die-Sintflut-in-Sage-Und-Wissenschaft.
2. François Lenormant, *Les Origines de l'Histoire d'après la Bible: de la Création de l'Homme au Déluge* (Paris, 1880).
3. Sir James G. Frazer, *The Great Flood: A Handbook of World Flood Myths* (CreateSpace Independent Publishing Platform, 2013), first published as chapter four of James George Frazer, *Folk-Lore in the Old Testament: Studies in Comparative Religion, Legend, and Law*, vol. 1 (Macmillan, 1923); Richard Andree, *Die Flutsagen* (F. Viewegundsohn, 1891); von Georg Gerland, *Der Mythus von der Sintflut* (Bonn, A. Marcus und E. Weber's Verlag, 1912).
4. Alan Dundes, ed., *The Flood Myth* (University of California Press, 1988), archive.org/details/floodmyth0000unse/page/n7/mode/2up.
5. Nick Liguori, *Echoes of Ararat: A Collection of Over 300 Flood Legends from North and South America* (Master Books, 2021).
6. Duane T. Gish, *Dinosaurs by Design* (Master Books, 1992), 74–75; John D. Morris, "Why Does Nearly Every Culture Have a Tradition of a Global Flood?," Institute for Creation Research, *Acts & Facts* 30, no. 9 (2001), icr.org/article/why-does-nearly-every-culture-have-tradition-globa/.
7. Flavius Josephus, *The Antiquities of the Jews* (94 AD), book 1, chap. 3, paragraph 6, archive.org/details/theAntiquitiesOfTheJews_507/page/n11/mode/2up.
8. John Woodward, *An Essay Toward a Natural History of the Earth and Terrestrial Bodies, Especially Minerals: As Also of the Sea, Rivers, and Springs, with an Account of the Universal Deluge; And the Effects That It Had upon the Earth* (Ric. Wilkin at the Kings-Head in St. Paul's Church-yard, 1695), 167–168, archive.org/details/essaytowardnatur00wood/page/n17/mode/2up.
9. Frazer, *The Great Flood*, 208–213.
10. Riem, *Die Sintflut in Sage und Wissenschaft*; Lenormant, *Les Origines de l'Histoire d'apres la Bible*; Frazer, *The Great Flood*; Dundes, *The Flood Myth*.
11. Charles Martin, *Flood Legends: Global Clues of a Common Event* (Master Books, 2009).
12. Byron C. Nelson, *The Deluge Story in Stone: A History of the Flood Theory of Geology* (Baker Book House, 1968), 166–167. This book is a reprint of the original published in 1931 by Augsburg Publishing House.
13. Francis A. Schaeffer, *Genesis in Space and Time: The Flow of Biblical History* (InterVarsity Press, 1972), 129–130.

Chapter 4: Which Flood Is *the* Flood?

1. Sir Leonard Woolley, *Ur of the Chaldees: A Record of Seven Years of Excavation* (Penguin Books, 1952, reprinted with revisions from the original in 1929), 15–25, archive.org/details/urofchaldeesreco00wool/page/14/mode/2up.
2. L. Charles Watelin and Stephen Langdon, *Excavations at Kish*, vol. 4 (Geuthner, 1934); J. L. Myres, "Excavations at Kish: Oxford Field-Museum Expedition," vol. 3, 1925–1927, by L. Ch. Watelin, with notes and selection of contracts by S. Langdon (Geuthner, 1930), 20, plates I–XVI; Henry Field, *The Field Museum—Oxford University Expedition to Kish, Mesopotamia, 1923–1929*, leaflet 28 of the Field Museum of Natural History (1929), 34, plates I–XIV; "The

Excavation at Kish," report of a lecture by Stephen Langdon, given to an unnamed society after the seventh season's work, reprint, 291–300; "Excavations at Kish 1928–1929," report of a lecture by Stephen Langdon, given to the Royal Asiatic Society (December 12, 1929), reprint from the Journal of the Royal Asiatic Society 1930, 601–610, with plates VII–XI, *Antiquity*, vol. 6, no. 23 (1932): 380–381, doi:10.1017/S0003598X00007250.

3. Erich Schmidt, "Excavations at Fara, 1931," *The Museum Journal*, Museum of the University of Pennsylvania, XXII, nos. 3–4 (1931): 193–245, penn.museum/documents/publications/journal/22-3and4/fara_excavations_1931.pdf.
4. William B. F. Ryan et al., "An Abrupt Drowning of the Black Sea Shelf," *Marine Geology* 138, nos. 1–2 (April 1997): 119–126, doi:10.1016/S0025-3227(97)00007-8.
5. William Ryan and Walter Pitman, *Noah's Flood: The New Scientific Discoveries About the Event That Changed History* (Simon & Schuster, 1998).
6. For a review of this evidence see R. D. Ballard, D. F. Coleman, and G. D. Rosenberg, "Further Evidence of Abrupt Holocene Drowning of the Black Sea Shelf," *Marine Geology* 170, nos. 3–4 (November 15, 2000): 253–261, doi:10.1016/S0025-3227(00)00108-0.
7. Liviu Giosan, Florin Filip, and Stefan Constatinescu, "Was the Black Sea Catastrophically Flooded in the Early Holocene?," *Quaternary Science Reviews* 28, nos. 1–2 (January 2009): 1–6, doi:10.1016/j.quascirev.2008.10.012.
8. Giosan, Filip, and Constatinescu, "Was the Black Sea," 1.
9. A. E. Aksu and R. N. Hiscott, "Persistent Holocene Outflow from the Black Sea to the Eastern Mediterranean Sea Still Contradicts the Noah's Flood Hypothesis: A Review of 1997–2021 Evidence and a Regional Paleoceanographic Synthesis for the Latest Pleistocene–Holocene," *Earth-Science Reviews* 227 (April 2022): id. 103960, doi:10.1016/j.earscirev.2022.103960.
10. Aksu and Hiscott, "Persistent Holocene Outflow," 1.
11. Aksu and Hiscott, 2.
12. Viacheslav K. Gusiakov et al., "Mega Tsunami of the World Oceans: Chevron Dune Formation, Micro-Ejecta, and Rapid Climate Change as the Evidence of Recent Oceanic Bolide Impacts," in *Geophysical Hazards: Minimizing Risk, Maximizing Awareness*, ed. Tom Beer (Springer, 2010), 197–227, doi:10.1007/978-90-481-3236-2.
13. Pierre Deschamps et al., "Ice Sheet Collapse and Sea-Level Rise at the Bølling Warming 14,600 Years Ago," *Nature* 483, no. 7391 (March 28, 2012): 559–564, doi:10.1038/nature10902; Till Hanebuth, Karl Stattegger, and Pieter M. Grootes, "Rapid Flooding of the Sunda Shelf: A Late-Glacial Sea Level Record," *Science* 288, no. 5468 (May 12, 2000): 1033–1035, doi:10.1126/science.288.5468.1033.
14. Dario Civile et al., "The Lost Adventure Archipelago (Sicilian Channel, Mediterranean Sea): Morphobathymetry and Late Quaternary Palaeogeographic Evolution," *Global and Planetary Change* 125 (February 2015): 36–47, doi:10.1016/j.gloplacha.2014.12.003.
15. Garry K. C. Clarke et al., "Paleohydraulics of the Last Outburst Flood from Glacial Lake Agassiz and the 8200 BP Cold Event," *Quaternary Science Reviews* 23, nos. 3–4 (February 2004): 389–407, doi:10.1016/j.quascirev.2003.06.004; Wallace S. Broecker, "Was the Younger Dryas Triggered by a Flood?," *Science* 312, no. 5777 (May 26, 2006): 1146–1148, doi:10.1126/science.1123253; Yong-Xiang Li et al., "Synchronizing a Sea-Level Jump, Final Lake Agassiz Drainage, and Abrupt Cooling 8200 Years Ago," *Earth and Planetary Science Letters* 315–316 (January 15, 2012): 41–50, doi:10.1016/j.epsl.2011.05.034; Timothy G. Fisher, Derald G. Smith, and John T. Andrews, "Preboreal Oscillation Caused by a Glacial Lake Agassiz Flood," *Quaternary Science Reviews* 21, nos. 8–9 (April 2002): 873–878, doi:10.1016/S0277-3791(01)00148-2; Chris S. M. Turney and Heidi Brown, "Catastrophic

Early Holocene Sea Level Rise, Human Migration and the Neolithic Transition in Europe," *Quaternary Science Reviews* 26, nos. 17–18 (September 2007): 2036–2041, doi:10.1016/j.quascirev.2007.07.003.

16. Turney and Brown, "Catastrophic Early Holocene Sea Level Rise," "'Noah's Flood' Kick-Started European Farming," *Phys Org News* (November 19, 2007), accessed June 5, 2023, phys.org/news/2007-11-noah-kick-started-european-farming.html.
17. Ryan and Pitman, "Noah's Flood," 138–139, 157–160.

Chapter 5: Before the Scientific Revolution

1. Terry Mortenson, "Does Belief in Inerrancy Necessitate a Particular View of the Age of the Earth?," *SES Steadfast Apologetics & Evangelism National Conference*, October 11, 2024, held at First Baptist Church of Rock Hill, Rock Hill, South Carolina.
2. Ian T. Taylor, *In the Minds of Men: Darwin and the New World Order* (TFE Publishing, 1984), 282–285; Mark A.Van Bebber and Paul S. Taylor, *Creation and Time: A Report on the Progressive Creationist Book by Hugh Ross* (Eden Communications, 1996), 93–104; Jonathan Sarfati, *Refuting Compromise: A Biblical and Scientific Refutation of "Progressive Creationism" (Billions of Years) As Popularized by Astronomer Hugh Ross* (Master Books, 2004), 107–125; James R. Mook, "The Early Church on Creation," *Answers Magazine*, August 20, 2015, 66–67, answersingenesis.org/church/the-early-church-on-creation/.
3. Justin Martyr, *Dialogue with Trypho*, chap. 138 in *Ante-Nicene Fathers*, vol. 1, ed. Alexander Roberts, James Donaldson, and A. Cleveland Coxe (Hendrickson Publishers, 1999), 268.
4. Martyr, *Dialogue with Trypho*, chap. 138, in *Ante-Nicene Fathers*, 268.
5. Theophilis, *Theophilis to Autolycus*, book 3, chap. 18, in *Ante-Nicene Fathers*, vol. 2, ed. Alexander Roberts, James Donaldson, and A. Cleveland Coxe (Hendrickson Publishers, 1999), 116.
6. Theophilis, *Theophilis to Autolycus*, book 3, chap. 19, in *Ante-Nicene Fathers*, 117.
7. Theophilis, *Theophilis to Autolycus*, book 3, chap. 19, in *Ante-Nicene Fathers*, 116.
8. Theophilis, *Theophilis to Autolycus*, book 2, chaps. 31–32, in *Ante-Nicene Fathers*, 106–107.
9. Irenaeus, *Against Heresies*, book 4, chap. 36, in *Ante-Nicene Fathers*, vol. 1, ed. Alexander Roberts, James Donaldson, and A. Cleveland Coxe (Hendrickson Publishers, 1999), 515–516.
10. Clement of Alexandria, *Stromata*, book 1, chap. 21 in *Ante-Nicene Fathers*, vol. 2, ed. Alexander Roberts, James Donaldson, and A. Cleveland Coxe (Hendrickson Publishers, 1999), 324.
11. Clement of Alexandria, *Stromata*, book 1, chap. 21 in *Ante-Nicene Fathers*, 331.
12. Tertullian, *On the Apparel of Women*, book 1, chap. 3 in *Ante-Nicene Fathers*, vol. 4, ed. Alexander Roberts, James Donaldson, and A. Cleveland Coxe (Hendrickson Publishers, 1999), 15.
13. Tertullian, *On the Pallium*, chap. 2 in *Ante-Nicene Fathers*, 6.
14. Tertullian, *The Treatise Against Hermogenes*, chap. 29, "The Gradual Development of Cosmical Order Out of Chaos in the Creation, Beautifully Stated," in *Ante-Nicene Fathers*, vol. 3, ed. Alexander Roberts, James Donaldson, and A. Cleveland Coxe (Hendrickson Publishers, 1999), 493–494.
15. Origen, *Origen Against Celsus*, book 6, chaps. 58–59 in *Ante-Nicene Fathers*, vol. 4, ed. Alexander Roberts, James Donaldson, and A. Cleveland Coxe (Hendrickson Publishers, 1999), 600.
16. Origen, *Origen Against Celsus*, book 6, chap. 59 in *Ante-Nicene Fathers*, 600.

17. Origen, *Homilies on Genesis*, Homily II, in *Homilies on Genesis and Exodus*, The Fathers of the Church: A New Translation, trans. Ronald E. Heine (Catholic University Press, 1982), 72–88, doi:10.2307/j.ctt32b3pv.
18. Gregory of Nyssa, *The Second Theological Oration, On Theology*, paragraph 18, in *Gregory of Nazianzus: Five Theological Orations*, trans. Stephen Reynolds (Estate of Stephen Reynolds, 2011), 27, accessed June 9, 2023, tspace.library.utoronto.ca/bitstream/1807/36303/1/Gregory%20of%20Nazianzus%20Theological%20Orations.pdf.
19. St. John Chrysostom, *Homilies on Genesis*, Homily 25, in *Homilies on Genesis 18–45, The Fathers of the Church: A New Translation*, trans. Robert C. Hill (Catholic University of America Press, 2010), 131.
20. Chrysostom, *Homilies on Genesis*, Homily 25, in *Homilies on Genesis*, 128–131.
21. Chrysostom, 128.
22. Chrysostom, 128–135.
23. Chrysostom, 140.
24. Chrysostom, 138.
25. Chrysostom, 138–139.
26. Jerome of Stricon, letter 51, section 5, in *Nicene and Post-Nicene Fathers*, vol. 6, *Jerome: Letters and Select Works*, second series, ed. Philip Schaff and Henry Wace, trans. W. H. Freemantle (Hendrickson, 1999), 87.
27. Jerome of Stricon, *Hebrew Questions on Genesis*, trans. C. T. R. Hayward (Clarendon Press, 1995), 37–38, isidore.co/CalibreLibrary/Jerome,%20St_/Hebrew%20Questions%20on%20Genesis%20(7397)/Hebrew%20Questions%20on%20Genesis%20-%20Jerome,%20St_.pdf.
28. Augustine, "The City of God," book 15, chap. 27, in *Nicene and Post-Nicene Fathers*, vol. 2, ed. Philip Schaff (Hendrickson Publishers, 1999), 307–308.
29. Augustine, "The Writings Against the Manichaeans, Reply to Faustus the Manichaean," book 12, chaps. 14–24 in *Nicene and Post-Nicene Fathers*, vol. 4, ed. Philip Schaff (Hendrickson Publishers, 1999), 188–191.
30. Augustine, "The Writings Against the Manichaeans," 188.
31. Augustine, 188.
32. Augustine, 188.
33. Augustine, 188.
34. Cyril of Alexandria, *Commentary on the Gospel According to John*, vol. 1, book 1, chap. 9, A Library of Fathers of the Holy Catholic Church, vol. 43, trans. P. E. Pusey (sold by James Parker and Rivington: Library of Fathers of the Holy Catholic Church, 1874), 98, accessed June 13, 2023, archive.org/details/CyrilOfAlexandriaCommentaryOnJohnVolume1Tr.P.E.Pusey1874, 152.
35. Theodoret of Cyrus, *The Questions on the Octateuch: On Genesis and Exodus*, vol. 1, Greek text rev. John F. Petruccione, trans. Robert C. Hill (Catholic University of America Press, 2007), 107–112, accessed June 12, 2023, archive.org/details/questionsonoctat0001theo/page/106/mode/2up?view=theater.
36. Saint Bede (the Venerable), *The Complete Works of Venerable Bede: Commentaries on the Scriptures*, Commentarii in Principium Genesis, ed. John Allen Giles (Whittaker, 1844), accessed June 28, 2023, play.google.com/books/reader?id=hYZPAAAAYAAJ&pg=GBS.PP14&hl=en.
37. Charles W. Jones, "Some Introductory Remarks on Bede's Commentary on Genesis," *Sacris Erudiri* 19 (1970): 169, doi:10.1484/J.SE.2.303290.
38. Bede the Venerable, *Translation of In Genesim I.1093-1224*, in James Aylett, trans., *Bede's*

Commentary on the Six Days of Creation and the Six Ages of the World, 27, accessed June 28, 2023, tartarus.org/verity/Bede_ug.pdf.
39. Bede the Venerable, *Translation of In Genesim I*, 27.
40. Bede the Venerable, 27.
41. Petrus Comestor, *Historia Scholastica: Eruditissimi viri magistri Petri Comestoris excellens opus quod Historia Scholastica inscribitur, magnam sacre scripture partem, que & in serie et in glossis crebro diffusa erat, breuiter complectens, mendis omnibus post omnes omnium hactenus editiones seclusis, in lucem exit*, chaps. xxxii–xxxv (National Central Library of Rome, 1543), 13–15 in the original book, 40–44 in the Google books online version, accessed June 28, 2023, archive.org/details/bub_gb_uUZfsY581hwC/page/n39/mode/2up.
42. Vincent de Beauvais, *Speculum Historiale* (September 5, 1494), accessed June 30, 2023, archive.org/details/A336104/mode/2up.
43. Thomas Aquinas, *Summa Theologiae*, part 1, question 1, article 10, reply to objection 1, 786, trans. Fathers of the English Dominican Province (1485), accessed June 30, 2023, archive.org/details/SummaTheologiae/page/n785/mode/2up.
44. Thomas Aquinas, *Commentary on the Book of Job*, ed. Joseph Kenny, trans. Brian Mulladay, accessed June 30, 2023, isidore.co/aquinas/english/SSJob.htm#0.
45. Aquinas, *Summa Theologica*, part 1, question 19, article 7, reply to objection 1, 148, accessed December 8, 2024, archive.org/details/SummaTheologicaThomasAquinas/page/n297/mode/2up.
46. Aquinas, *Summa Theologica*, part 3, question 46, article 4, seventh reason, 3020, accessed December 8, 2024, archive.org/details/SummaTheologicaThomasAquinas/page/n3169/mode/2up.
47. Aquinas, *Summa Theologica*, part 3, question 66, article 11, reply to objection 3, 3185, accessed December 8, 2024, archive.org/details/SummaTheologicaThomasAguinas/page/n3335/mode/2up.
48. Nicholas of Lyra, *Postilla Litteralis in Vetus et Novum Testamentum* (1432), 45–47, accessed June 30, 2023, archive.org/details/A332145/page/n45/mode/2up.
49. Martin Luther, *Commentary on Genesis: Creation, Sin, and the Flood*, trans. John Nicholas Lenker (Lutheran Library Publishing, 2021), accessed July 1, 2023, lutheranlibrary.org/pdf/356-luther-genesis.pdf.
50. Luther, *Commentary on Genesis*, chap. 6, paragraphs 197–207, 561–564.
51. Luther, chap. 6, paragraph 213, 565.
52. Luther, paragraph 218, 566.
53. Luther, paragraph 217, 566.
54. Luther, paragraph 216, 566.
55. Luther, paragraph 230, 569.
56. Luther, paragraph 236, 571.
57. Luther, paragraph 217, 571.
58. Luther, paragraph 240, 572
59. Luther, paragraph 238, 571.
60. Luther, paragraph 238, 572.
61. Luther, chap. 7, paragraph 69, 604.
62. Luther, paragraph 65, 603.
63. Luther, chap. 6, paragraphs 242–256, 573–576.
64. Luther, paragraphs 249–250, 574.
65. Luther, chap. 7, paragraphs 45–54, 598–600.

66. John Calvin, *Commentaries on the First Book of Moses Called Genesis*, vol. 1, trans. John King (Christian Classics Ethereal Library, 1847 from the original in 1563), chap. 6, verse 3, accessed July 2, 2023, ccel.org/ccel/calvin/calcom01.all.html.
67. Calvin, *Commentaries on the First Book*, chap. 6, verse 5.
68. Calvin, verse 7.
69. Calvin, verse 17.
70. Calvin, verse 17.
71. Calvin, chap. 7, verse 17.
72. John Calvin, Commentary on 2 Peter, accessed July 2, 2023, studylight.org/commentaries/eng/cal/2-peter-3.html.
73. Calvin, *Commentaries on the First Book*, chap. 7, verse 17.
74. Calvin, verse 17.
75. Calvin, chap. 8, verse 3.
76. Calvin, chap. 9, verse 13.
77. Calvin, verse 13.
78. Calvin, verse 13.

Chapter 6: The Debate Begins

1. Athanasius Kircher, *Arca Noë* (Johannes van Waesbergen, 1675), accessed July 17, 2023, gallica.bnf.fr/ark:/12148/bpt6k111864r.texteImage.
2. Kircher, *Arca Noë*, 16–45.
3. Kircher, 46–116.
4. Kircher, 117–124.
5. Kircher, 187–196.
6. Isaac Vossius, *De Septuaginta Interpretibus, eorumque Tralatione & Chronologia Dissertationes* (Adriani Vlacq, 1661), 279–284, 339–344, play.google.com/books/reader?id=9vqfnyBEkHUC&pg=GBS.PP24&hl=en.
7. Scott Mandelbrote, "Isaac Vossius and the Septuagint," in *Isaac Vossius (1618–1689) Between Science and Scholarship*, ed. Eric Jorink and Dirk Van Miert (Koninklijke Brill NV, 2012), 96–97.
8. Edward Stillingfleet, *Origins Sacrae, or a Rational Account of the Grounds of Christian Faith, as to the Truth and Divine Authority of the Scriptures, and the Matters Therein Contained*, 3rd ed., Of the Origins of Nations (Henry Martlock, 1666), book 3, chap. 4, 533, books.google.com/books?id=YGD3Cq598bIC&printsec=frontcover&source=gbs_ge_summary_r&cad=0#v=onepage&q&f=false.
9. Stillingfleet, *Origins Sacrae*, book 3, chap. 4, 533.
10. Stillingfleet, 538.
11. Stillingfleet, 538–539.
12. Stillingfleet, 539, 551–552; Genesis 7:10–11, 8:14–15.
13. Stillingfleet, 539–541.
14. Stillingfleet, 539.
15. Stillingfleet, 539.
16. Stillingfleet, 540.
17. Stillingfleet, 542.
18. Stillingfleet, 542–543.
19. Thomas Burnet, *The Sacred Theory of the Earth*, 2nd ed. (Centaur Press, 1965, R. Norton for Walter Kettilby, 1691), first book, chap. 2, 27, accessed July 13, 2023, archive.org/details/

sacredtheoryofea1965burn/page/26/mode/2up.
20. Burnet, *The Sacred History of the Earth*, first book, chap. 2, 28–33.
21. Burnet, chap. 5, 53.
22. Burnet, chap. 3, 41; chap. 6, 64–69.
23. Burnet, chap. 6, 70.
24. John Woodward, *An Essay Toward a Natural History of the Earth and Terrestrial Bodies, Especially Minerals: As Also of the Sea, Rivers, and Springs, with an Account of the Universal Deluge; And the Effects That It Had upon the Earth* (Ric. Wilkin at the Kings-Head in St. Paul's Church-yard, 1695), archive.org/details/essaytowardnatur00wood/page/n17/mode/2up.
25. Woodward, *An Essay Toward a Natural History*, part 3, 162.
26. Woodward, part 2, 74.
27. Woodward, part 3, 164.
28. Woodward, part 2, 73–83.
29. Woodward, 82.
30. Woodward, part 3, 165.
31. Woodward, part 4, 180–197.
32. Woodward, part 2, 107–109, 117–118; part 3, 163–165; part 5, 232–233.
33. Woodward, part 2, 108.
34. Woodward, part 4, 199.
35. Woodward, part 3, 134–136, 140; part 4, 189, 199.
36. Woodward, part 2, 81.
37. Woodward, 75–77.
38. Woodward, 77–78.
39. Woodward, 84.
40. Woodward, 85–90.
41. Woodward, part 6, 266.
42. Woodward, part 2, 84.
43. Woodward, 85.
44. Woodward, 85.
45. Woodward, 105.
46. Woodward, 86.
47. Woodward, 94–95.
48. Woodward, 92.
49. Edmond Halley, "VII. Some Considerations About the Cause of the Universal Deluge, Laid Before the Royal Society, on the 12th of December 1694," *Philosophical Transactions of the Royal Society* 33, no. 383 (June 30, 1724): 118–123, doi:10.1098/rstl.1724.0023; Edmond Halley, "VIII. Some Farther Thoughts upon the Same Subject Delivered on the 19th of the Same Month," *Philosophical Transactions of the Royal Society* 33, no. 383 (June 30, 1724): 123–125, doi:10.1098/rstl.1724.0024.
50. Halley, "VII. Some Considerations About the Cause," 119–120.
51. Halley, 121.
52. Halley, 121.
53. Halley, 121.
54. Halley, 118.
55. William Whiston, *A New Theory of the Earth* (R. Roberts for B. Tooke, 1696), book 2, Hypotheses, section 3, 97–99, book 3, chap. 2, sections 28–32, 172–173, accessed July 13, 2023, archive.org/details/newtheoryofearth0000whis/page/98/mode/2up.

56. Whiston, *New Theory of the Earth*, book 2, Hypotheses, section 8, 119; book 4, chap. 4, section 56, 319.
57. Whiston, book 2, Scholium, 154–156, book 3, chap. 4, sections 45–53, 188–192; book 4, chap. 4, sections 49–58, 309–323.
58. Whiston, book 3, chap. 4, sections 48–53, 189–192.
59. Whiston, book 3, chap. 4, section 57, 193.
60. Whiston, book 3, chap. 4, sections 54–56, 192–193; book 4, chap. 4, section 54, 316.
61. Matthew Henry, *An Exposition of the Old and New Testaments*, 1st American ed., vol. 1 (Ed. Barrington & Geo. D. Baswell, 1828), 60–78, archive.org/details/expositionofoldn01henr/page/n7/mode/2up; biblestudytools.com/commentaries/matthew-henry-complete/.
62. Henry, *An Exposition of the Old and New Testaments*, 62.
63. Henry, 63–66, 69.
64. Henry, 64–65.
65. Henry, 67.
66. Henry, 67.
67. Henry, 68.
68. Henry, 68.
69. John Wesley, *Explanatory Notes on the Bible: Notes on the First Book of Moses Called Genesis* (Wesley Center Online, Wesleyan Heritage Publishing), chap. 6, wesley.nnu.edu/john-wesley/john-wesleys-notes-on-the-bible/notes-on-the-first-book-of-moses-called-genesis/#Chapter%2BVI.
70. Wesley, *Explanatory Notes on the Bible: Genesis*, chap. 6.
71. Wesley, chap. 6.
72. Wesley, chap. 7.
73. Wesley, chap. 7.
74. Wesley, chap. 7.
75. Wesley, chap. 8.
76. Edward Neville Da Costa Andrade, "Wilkins Lecture—Robert Hooke," *Proceedings of the Royal Society A* 201, no. 1067 (May 23, 1950): 466, doi:10.1098/rspa.1950.0071; Ellen T. Drake and Paul D. Komar, "A Comparison of the Geological Contributions of Nicolaus Steno and Robert Hooke," *Journal of Geological Education* 29, no. 3 (August 1981): 127–134, doi:10.5408/0022-1368-29.3.127.
77. Andrade, "Wilkins Lecture," 467–468.
78. Nicolaus Steno, *De Solido intra Solidum Naturaliter Contento Dissertationis Prodromos* (Ex typographiae sub signo Stellae, 1669), 52–76, accessed August 7, 2023, archive.org/details/nicolaistenonisd00sten/page/12/mode/2up.
79. James Hutton, "Theory of the Earth; or an Investigation of the Laws Observable in the Composition, Dissolution, and Restoration of Land upon the Globe," *Transactions of the Royal Society of Edinburgh* vol. 1, part 2 (1788), 209–304, accessed August 7, 2023, archive.org/details/cbarchive_106252_theoryoftheearthoraninvestigat1788/page/n1/mode/2up.
80. William Buckland, *Vindiciae Geologicae; or the Connexion of Geology with Religion Explained, in an Inaugural Lecture Delivered Before the University of Oxford, May 15, 1819, on the Endowment of a Readership in Geology by His Royal Highness the Prince Regent* (Oxford University Press, 1820), 23–24, accessed August 7, 2023, archive.org/details/b22393304.
81. Georges Cuvier, *Essay on the Theory of the Earth*, 3rd ed. (William Blackwood, 1817), section 34, 171, accessed August 10, 2023, archive.org/details/b21299961.
82. Buckland, *Vindiciae Geologicae*, 4–33.

83. Buckland, 30.
84. William Buckland, *The Bridgewater Treatises: Treatise VI, On Geology and Mineralogy*, vol. 1 (William Pickering, 1837), chap. 1, 1–8; chaps. 3–12, 34–128, books.google.com/books?id=MVoOAAAAQAAJ&printsec=frontcover#v=onepage&q&f=false.
85. Buckland, *Bridgewater Treatise VI*, chap. 2, 16–17; chap. 9, 94–95.
86. Charles Lyell, *Principles of Geology: Being an Attempt to Explain the Former Changes of the Earth's Surface by Reference to Causes Now in Operation,* vol. 3 (John Murray, 1833), 8–389, darwin-online.org.uk/content/frameset?viewtype=text&itemID=A505.3&pageseq=1; Sir Charles Lyell, *Principles of Geology. Or, the Modern Changes of the Earth and Its Inhabitants Considered as Illustrative of Geology*, 9th ed. (D. Appleton, 1854), 61–775, gutenberg.org/cache/epub/33224/pg33224-images.html.
87. Michael Pidwirny and Michael Scott Jones, "Introduction to the Lithosphere," in *Fundamentals of Physical Geography*, 2nd ed. (University of British Columbia, Okanagan, 1999), chap. 10, section C: Concept of Uniformitarianism, accessed April 16, 2024, physicalgeography.net/fundamentals/10c.html.
88. Adam Sedgwick, "Address to the Geological Society, Delivered on the Evening of the 18th of February 1831," *Proceedings of the Geological Society of London* 1 (1834): 313. This statement from Sedgwick's address to the Geological Society is also available online at John Willis Clark and Thomas McKenny Hughes, *The Life and Letters of the Reverend Adam Sedgwick*, vol. 1 (Cambridge University Press, 1890), 370, accessed August 10, 2023, archive.org/details/in.ernet.dli.2015.204554/page/n393/mode/2up.
89. George Bugg, *Scriptural Geology, or, Geological Phenomena, Consistent with the Literal Interpretation of the Sacred Scriptures, Upon the Subjects of the Creation and Deluge*, vol. 1 (Hatchard and Son, 1826); George Bugg, *Scriptural Geology, or, Geological Phenomena, Consistent with the Literal Interpretation of the Sacred Scriptures*, vol. 2 (Hatchard and Son, 1827); George D. D. Young, *Scriptural Geology, or, an Essay on the High Antiquity Ascribed to the Organic Remains Imbedded in Stratified Rocks* (Simpkin, Marshall, and Co., 1838).
90. In a lengthy review of the history of scriptural geologists, Warren Johns names a total of twenty-five that comprised the group. See Warren H. Johns, "Scriptural Geology, Then and Now," *Answers Research Journal* 9 (2016): 317–337, accessed August 11, 2023, answersresearchjournal.org/scriptural-geology-then-and-now/.
91. David N. Lord, *Geognosy: Facts and Principles of Geology Against Theories* (Franklin Knight, 1855), 45, archive.org/details/geognosy00lord/page/44/mode/2up.
92. William Buckland, *Bridgewater Treatise VI. Geology and Mineralogy Considered with Reference to Natural Theology*, vol. 1 (William Pickering, 1837), chap. 2, 14, accessed August 10, 2023, archive.org/details/bridgewatertreat61chal/page/8/mode/2up.
93. Bugg, *Scriptural Geology*, 124; Biblicus Delvinus (pseudonym of George Bugg), *A Brief Treatise on Geology; or Facts, Suggestions, and Inductions in That Science*, 2nd ed. (L. and G. Seeley; Hatchard and Son, 1839), 25–33, 40–44, accessed August 11, 2023, books.google.je/books?id=H91JAAAAIAAJ&printsec=frontcover&output=html_text&source=gbs_book_other_versions_r&cad=1.
94. Philip Henry Gosse, *Omphalos: An Attempt to Untie the Geological Knot* (John Van Voorst, 1857).
95. Charles Darwin, *On the Origin of Species: By Means of Natural Selection, or the Preservation of Favoured Races in the Struggle for Life* (John Murray, 1859).

Chapter 7: Controversy Reignited

1. Jerry A. Coyne, *Faith vs. Fact: Why Science and Religion Are Incompatible* (Penguin Books, 2016), 14.
2. Coyne, *Faith vs. Fact*, 2.
3. Coyne, 2.
4. Charles Darwin, *The Descent of Man, and Selection in Relation to Sex*, vols. 1 and 2 (John Murray, 1871).
5. Charles Darwin, *The Descent of Man, and Selection in Relation to Sex* in *From So Simple a Beginning: Darwin's Four Great Books*, ed. Edward O. Wilson (W. W. Norton and Co., 2006), 1248.
6. Darwin, *The Descent of Man*, in *From So Simple a Beginning*, 1239.
7. Rev. P. R. Russel, "Darwinism Examined," *The Advent Review and Sabbath Herald* 47, no. 20 (May 18, 1876): 153, documents.adventistarchives.org/Periodicals/RH/RH18760518-V47-20.pdf.
8. Frank M. Hasel, "Ellen G. White and Creationism: How to Deal with Her Statements on Creation and Evolution: Implications and Prospects," *Journal of the Adventist Theological Society* 17, no. 1 (Spring 2006): 231–232.
9. Ellen G. White, *Spiritual Gifts*, vol. 3 (Seventh-Day Adventist Publishing, 1864), 90, accessed August 21, 2023, next.beta.egwwritings.org/book/b106.
10. E. G. White, *Noah: Another Storm is Coming* (Remnant Publications, 2014), 37–38, archive.org/details/noahanotherstorm0000whit/page/36/mode/2up.
11. White, *Noah*, 38.
12. White, 43.
13. White, 45.
14. George McCready Price, *Illogical Geology: The Weakest Point in the Evolution Theory* (Modern Heretic Company, 1906), accessed August 21, 2023, play.google.com/books/reader?id=d4oNAAAAYAAJ&pg=GBS.PP2&hl=en.
15. Price, *Illogical Geology*, 7.
16. Price, 9.
17. George McCready Price, *The Fundamentals of Geology and Their Bearings on the Doctrine of a Literal Creation* (Pacific Press Publishing, 1913); George McCready Price, *The New Geology* (Pacific Press Publishing Association, 1923).
18. Harry Rimmer, *Modern Science, Noah's Ark, and the Deluge* (Research Science Bureau, 1925), 28.
19. John C. Whitcomb and Henry M. Morris, *The Genesis Flood: The Biblical Record and Its Scientific Implications* (Presbyterian & Reformed Publishing, 1961).
20. Henry M. Morris II, "Scientific Defense of Flood Geology and a Young Earth," lecture presented to the Reasons to Believe staff and volunteer apologists with the research staff of the Institute of Creation Research in attendance (1993).
21. Ronald L. Numbers, *The Creationists: The Evolution of Scientific Creationism* (Alfred A. Knopf, 1992), xi.
22. John R. Baumgardner, "Computer Modeling of the Large-Scale Tectonics Associated with the Genesis Flood," *Proceedings of the Third International Conference on Creationism*, vol. 3, no. 15, ed. R. E. Walsh (Creation Science Fellowship, 1994): 49–62; Steven A. Austin et al., "Catastrophic Plate Tectonics: A Global Flood Model of Earth History," *Proceedings of the Third International Conference on Creationism*, ed. R. E. Walsh (Creation Science Fellowship 1994): 609–621.

23. Larry Vardiman, Andrew A. Snelling, and Eugene F. Chaffin, eds., *Radioisotopes and the Age of the Earth: A Young-Earth Creationist Research Initiative*, vol. 1 (Institute for Creation Research; Creation Research Society, 2000); Larry Vardiman, Andrew A. Snelling, and Eugene F. Chaffin, eds., *Radioisotopes and the Age of the Earth: A Young-Earth Creationist Research Initiative*, vol. 2 (Institute for Creation Research; Creation Research Society, 2005). Also available online at icr.org/i/pdf/research/rate-all.pdf; icr.org/i/pdf/technical/RATE2-Preliminaries.pdf.
24. M. Santosh and D. I. Groves, "Global Metallogeny in Relation to Secular Evolution of the Earth and Supercontinent Cycles," *Gondwana Research* 107 (July 2022): 395–422, doi:10.1016/j.gr.2022.04.007; Zheng-Xiang Li, Yebo Liu, and Richard Ernst, "A Dynamic 2000–540 Ma Earth History: From Cratonic Amalgamation to the Age of Supercontinent Cycle," *Earth Science Reviews* 238 (March 2023): id. 104336, doi:10.1016/j.earscirev.2023.104336; Chuan Huang and Zheng-Xiang Li, "Why Supercontinents Became Shorter Lived as the Earth Evolved," *Science Bulletin* 68, no. 4 (February 2023): 436–440, doi:10.1016/j.scib.2023.01.035.
25. William D. Barrick, "Noah's Flood and Its Geological Implications," in *Coming to Grips with Genesis: Biblical Authority and the Age of the Earth*, ed. Terry Mortenson and Thane H. Ury (Master Books, 2008), 275–281; John R. Baumgardner, "Catastrophic Plate Tectonics: The Physics Behind the Genesis Flood," *Proceedings of the Fifth International Conference on Creationism*, vol. 5, no. 13 (Creation Science Fellowship, 2003), 113–126, digitalcommons.cedarville.edu/icc_proceedings/vol5/iss1/13; John Baumgardner, "Catastrophic Plate Tectonics: The Geophysical Context of the Genesis Flood," *Journal of Creation* 16, no. 1 (April 2002): 58–63, creation.com/catastrophic-plate-tectonics-the-geophysical-context-of-the-genesis-flood; Steven A. Austin et al., "Catastrophic Plate Tectonics: A Global Flood Model of Earth History," *Proceedings of the Third International Conference on Creationism*, vol. 3, no. 56, held July 18–23, 1994 (Creation Science Fellowship, 1994): 609–621, digitalcommons.cedarville.edu/icc_proceedings/vol3/iss1/56/.
26. Timothy D. Herbert, "The Mid-Pleistocene Climate Transition," *Annual Review of Earth and Planetary Sciences* 51 (May 2023): 389–418, doi:10.1146/annurev-earth-032320-104209; Richard A. Muller and Gordon J. MacDonald, "Glacial Cycles and Orbital Inclination," *Nature* 377 (September 14, 1995): 107–108, doi:10.1038/377107b0; Hugh Ross, *Weathering Climate Change: A Fresh Approach* (RTB Press, 2020), 68–79, 129–147.
27. Michael Oard, "The Ice Age and the Genesis Flood," *Acts & Facts: Impact* 16, no. 6 (June 1, 1987), accessed August 26, 2023, icr.org/article/ice-age-genesis-flood; John D. Morris, "The Ice Age: Causes and Consequences," *Acts & Facts: Back to Genesis* 38, no. 8 (August 1, 2009): 15, accessed August 26, 2023, icr.org/article/ice-age-causes-consequences; Andrew A. Snelling and Mike Matthews, "When Was the Ice Age in Biblical History?," *Answers Magazine*, April 1, 2013, 46–52, accessed August 26, 2023, answersingenesis.org/environmental-science/ice-age/when-was-the-ice-age-in-biblical-history/; Larry Vardiman, "What Started the Ice Age?," *Answers Magazine*, April 1, 2013, 53–55, accessed August 26, 2023, answersingenesis.org/environmental-science/ice-age/what-started-the-ice-age/.
28. Snelling and Matthews, "When Was the Ice Age in Biblical History?"; Vardiman, "What Started the Ice Age?"; Morris, "The Ice Age: Causes and Consequences."
29. Morris, "The Ice Age: Causes and Consequences"; Vardiman, "What Started the Ice Age?"
30. Terry Mortenson et al., "Affirmations and Denials Essential to a Consistent Christian (Biblical) Worldview," in *Coming to Grips with Genesis: Biblical Authority and the Age of the Earth*, ed. Terry Mortenson and Thane H. Ury (Master Books, 2008), 454, 456.

31. For example, when astronomers take a spectrum of the light from a star or galaxy, that spectrum yields a measurement of the laws and constants of physics when light left that star or galaxy. Thousands of such measurements establish that the laws and constants of physics from 4.2 years ago to 12.9 billion years ago (the light-travel times) have not changed by any more than 1 part in 10 quadrillion per year. For details and documentation see, Hugh Ross, "Bible Got It Right: Immutable Physical Laws," *Today's New Reason to Believe* (blog), Reasons to Believe, January 29, 2024, reasons.org/explore/blogs/todays-new-reason-to-believe/bible-got-it-right-immutable-physical-laws; Michael T. Murphy et al., "Fundamental Physics with ESPRESSO: Precise Limit on Variations in the Fine-Structure Constant Towards the Bright Quasar HE 0515-4414," *Astronomy & Astrophysics* 658 (February 2022): id. A123, doi:10.1051/0004-6361/202142257; T. D. Le, "Stringent Limit on Space-Time Variation of Fine-Structure Constant Using High-Resolution of Quasar Spectra," *Heliyon* 6, no. 9 (September 23, 2020): id. e05011, doi:10.1016/j.heliyon.2020.e05011; Haoran Liang and Zhe Wu, "Measuring the Fine-Structure Constant on Quasar Spectra: High Spectral Resolution Gains More Than Large Size of Moderate Spectral Resolution Spectra," *Proceedings of the SPIE 12644, International Workshop on Frontiers of Graphics and Image Processing, FGIP 2022*, vol. 12644 (May 3, 2023): id. 1264408, doi:10.1117/12.2670012; S. A. Levshakov et al., "An Upper Limit to the Variation in the Fundamental Constants at Redshift $z = 5.2$," *Astronomy & Astrophysics: Letters* 540 (April 2012): id. L9, doi:10.1051/0004-6361/201219042; Julija Bagdonaite et al., "A Stringent Limit on a Drifting Proton-to-Electron Mass Ratio from Alcohol in the Early Universe," *Science* 339, no. 6115 (January 4, 2013): 46–48, doi:10.1126/science.1224898; Earl Patrick Bellinger and Jørgen Christensen-Dalsgaard, "Asteroseismic Constraints on the Cosmic-Time Variation of the Gravitational Constant from an Ancient Main-Sequence Star," *Astrophysical Journal Letters* 887, no. 1 (December 3, 2019): id. L1, doi:10.3847/2041-8213/ab43e7.

Chapter 8: Hyperbole, Allegory, Myth, or What?

1. A few examples include Tremper Longman III and John H. Walton, *The Lost World of the Flood: Mythology, Theology, and the Deluge Debate* (IVP Academic, 2018); William Ryan and Walter Pitman, *Noah's Flood: The New Scientific Discoveries About the Event That Changed History* (Simon & Schuster, 1998); Nick Thom, *The Great Flood: The True Story of the Most Dramatic Event in All of Human History* (Grosvenor House Publishing, 2011).
2. Carol A. Hill, "The Noachian Flood: Universal or Local?," *Perspectives on Science and Christian Faith* 54, no. 3 (September 2002): 181, accessed June 2, 2023, asa3.org/ASA/PSCF/2002/PSCF9 02Hill.pdf.
3. Lorence G. Collins, "Yes, Noah's Flood May Have Happened, But Not Over the Whole Earth," *Reports of the National Center for Science Education* 29, no. 5 (September–October 2009), accessed June 2, 2023, ncse.ngo/yes-noahs-flood-may-have-happened-not-over-whole-earth.
4. Biblical Archaeology Society Staff, "Noah and the Flood," *Bible History Daily* (May 5, 2019), accessed June 2, 2023, biblicalarchaeology.org/daily/archaeology-today/biblical-archaeology-topics/noah-and-the-flood/.
5. Longman and Walton, *The Lost World of the Flood*, 36–41.
6. Longman and Walton, 38–40.
7. Longman and Walton, 41.
8. Longman and Walton, 38–40.
9. Longman and Walton, 22.

10. Israel Finkelstein and Neil Asher Silberman, *The Bible Unearthed: Archaeology's New Vision of Ancient Israel and the Origin of Its Sacred Texts* (Free Press, 2002).
11. Norman Cohn, *Noah's Flood: The Genesis Story in Western Thought* (Yale University Press, 1996).
12. William Lane Craig, "The Historical Adam," *First Things*, October 1, 2021, 2, accessed February 21, 2025, firstthings.com/article/2021/10/the-historical-adam.
13. William Lane Craig, *In Quest of the Historical Adam: A Biblical and Scientific Exploration* (Eerdmans, 2021), 127.
14. Craig, *In Quest of the Historical Adam*, 131.
15. Craig, 120–129.
16. Peter Enns, *Inspiration and Incarnation: Evangelicals and the Problem of the Old Testament*, 2nd ed. (Baker Academic, 2015), 39, 50.
17. Michael J. Kruger, "The Bible Tells Me So," *The Gospel Coalition, U.S. Edition: Bible & Theology* (October 17, 2014), accessed September 11, 2023, thegospelcoalition.org/reviews/bible-tells/.
18. Kruger, "The Bible Tells Me So."
19. Peter Enns, *The Bible Tells Me So: Why Defending Scripture Has Made Us Unable to Read It* (HarperOne, 2014), 128.
20. Enns, *Inspiration and Incarnation*, 39.
21. Peter Enns, *The Evolution of Adam: What the Bible Does and Doesn't Say About Human Origins* (Brazos Press, 2012), 50.
22. Enns, *The Evolution of Adam*, 50.
23. Enns, 58.
24. Enns, 56.
25. Kenton L. Sparks, *God's Word in Human Words: An Evangelical Appropriation of Critical Biblical Scholarship* (Baker Academic, 2008), 97–99.
26. Sparks, *God's Word in Human Words*, 98.
27. Alan Dundes, ed., *The Flood Myth* (University of California Press, 1988).
28. Alan Dundes, "The Flood as Male Myth of Creation," in *The Flood Myth*, ed. Alan Dundes (University of California Press, 1988), 167–181.
29. Louis Ginzberg, "Noah and the Flood in Jewish Legend," in *The Flood Myth*, ed. Alan Dundes (University of California Press, 1988), 319–335.
30. Louis Ginzberg, *The Legends of the Jews*, trans. Henrietta Szold (Jewish Publication Society of America, 1913), 145–181, accessed September 28, 2023, archive.org/details/legendsofjews01ginz/page/144/mode/2up.
31. John H. Walton and D. Brent Sandy, *The Lost World of Scripture: Ancient Literary Culture and Biblical Authority* (IVP Academic, 2013), 205–213.
32. Walton and Sandy, *The Lost World of Scripture*, 205.
33. Enns, *Inspiration and Incarnation*, 27–28, 44, 56, 173–177; Sparks, *God's Word in Human Words*, 55, 230–231, 242–259.
34. R. Laird Harris, Gleason L. Archer Jr., and Bruce K. Waltke, *Theological Wordbook of the Old Testament*, vol. 1 (Moody Press, 1980), 378–379.
35. Walter C. Kaiser Jr., "The Literary Form of Genesis 1–11," in *New Perspectives on the Old Testament*, ed. J. Barton Payne (Word Books, 1970), 59.
36. Kaiser, "The Literary Form of Genesis 1–11," 59–60.
37. Longman and Walton, *The Lost World of the Flood*, 11.
38. Enns, *The Evolution of Adam*, xiv, 4, 79; Enns, *Inspiration and Incarnation*, 16–17, 29–30,

43–44; Sparks, *God's Word in Human Words*, 12–13, 83–84.
39. Sparks, *God's Word in Human Words*, 256.

Chapter 9: The Genesis Flood and the Mesopotamian Epics

1. Thorkild Jacobsen, "The Eridu Genesis," in *I Studied Inscriptions from Before the Flood: Ancient Near Eastern, Literary, and Linguistic Approaches to Genesis 1–11*, ed. Richard S. Hess and David Toshio Tsumura (Eisenbrauns, 1994), 129, accessed October 5, 2023, books.google.com/books?id=g5MGVP6gAPkC&q=Eridu+Genesis.+Nippur&pg=PA129#v=snippet&q=Eridu%20Genesis.%20Nippur&f=false.
2. W. G. Lambert and A. R. Millard, *Atra-Hasīs: The Babylonian Story of the Flood* (Eisenbrauns, 1999). A free online translation of the key elements of *The Epic of Atrahasis* is available at livius.org/sources/content/anet/104-106-the-epic-of-atrahasis/. Additional elements are availableat omnika.org/texts/66.
3. Maureen Gallery Kovacs, trans., *The Epic of Gilgamesh: Tablet XI The Story of the Flood*, (1998), accessed October 6, 2023, ancienttexts.org/library/mesopotamian/gilgamesh/tab11.htm.
4. Kovacs, *The Epic of Gilgamesh*, lines 82–83.
5. Kovacs, lines 65–72, 78–79, 85–90.
6. Kovacs, line 92.
7. Kovacs, line 113.
8. Kovacs, line 115.
9. Kovacs, lines 105–106.
10. Kovacs, lines 107–112.
11. Kovacs, line 128.
12. Kovacs, line 134.
13. Kovacs, lines 140–144.
14. Kovacs, line 156.
15. Kovacs, line 297.
16. Kovacs, line 301.
17. Kovacs, lines 307–308.
18. Pete Enns, "Gilgamesh, Atrahasis and the Flood," *BioLogos, The Bible in Ancient Context, The Flood* (June 1, 2010), accessed October 6, 2023, biologos.org/articles/gilgamesh-atrahasis-and-the-flood.
19. Ruth E. Simoons-Vermeer, "The Mesopotamian Floodstories: A Comparison and Interpretation," *Numen* 21, no. 1 (April 1974): 17–34, doi:10.2307/3269712.
20. E. A. Speiser trans., "Akkadian Myths and Epics" in *Ancient Near Eastern Texts Relating to the Old Testament, Third Edition with Supplement*, ed. James B. Pritchard (Princeton University Press, 1969), 69–119, 503–517, accessed January 2, 2025; S. N. Kramer, trans., "Sumerian Myths and Epic Tales," in *Ancient Near Eastern Texts Relating to the Old Testament, Third Edition with Supplement*, ed. James B. Pritchard (Princeton University Press, 1969), 37–58, accessed January 2, 2025, tarsus.ie/resources/Wisdom-Lit.-/ANET-PDF.pdf.
21. Many missionaries working with modern-day stone-age tribes have observed these amazing memorization skills. See, for example, Don Richardson, *Eternity in Their Hearts* (Regal Books, 1984).
22. Meredith G. Kline, "Job: Introduction," in *The Wycliffe Bible Commentary: A Phrase by Phrase Commentary of the Bible*, ed. Charles F. Pfeiffer and Everett F. Harrison (Moody Bible Institute of Chicago, 1962), 459; John Franklin Genung, "Book of Job: Setting of Time,

Place, and Scene," in *The International Standard Bible Encyclopaedia*, vol. 3: Heresy–Peleg, ed. James Orr et al. (Wm. B. Eerdmans, 1956), 1680.

Chapter 10: God's Response to Reprobation

1. WestminsterStandards.org, *Westminster Confession of Faith, chapter II. Of God, and of the Holy Trinity,* westminsterstandards.org/westminster-confession-of-faith/.
2. Christian Reformed Church, *Ecumenical Creeds and Reformed Confessions* (CRC Publications, 1987), 78.
3. Tim Keller, *Forgive: Why Should I and How Can I?* (Viking, 2022), 74.
4. Keller, *Forgive*, 74.
5. Matthew 5:18–20; James 2:10.
6. Matthew 5:21–22.
7. Romans 5:12–14; Jeremiah 17:9–10.
8. Psalm 51:5.
9. Job 4:8; Proverbs 5:22, 11:5; Isaiah 57:20–21; Jeremiah 5:25; Romans 7:5; Galatians 5:19–21, 6:7–8.
10. 1 Corinthians 6:12–20.
11. 1 Corinthians 6:18–19.
12. R. Laird Harris, Gleason L. Archer Jr., and Bruce K. Waltke, *Theological Wordbook of the Old Testament*, vol. 2 (Moody Press, 1980), 587–591.
13. Ephesians 1:13–14.
14. 2 Peter 2:14, 18–19.
15. Romans 6:16; 2 Timothy 2:26.
16. Romans 1:32.
17. Harris, Archer, and Waltke, *Theological Wordbook of the Old Testament*, vol. 2, 917–918; William Gesenius, *Gesenius' Hebrew-Chaldee Lexicon to the Old Testament* (Baker, 1979), 815–816.
18. Harris, Archer, and Waltke, *Theological Wordbook of the Old Testament*, vol. 1, 297; Gesenius, *Hebrew-Chaldee Lexicon*, 288.
19. Matthew 24:9–12; Revelation 9:20–21, 13:5–8, 16:10–11.
20. Genesis 4:23–24.
21. Hugh Ross, *Navigating Genesis: A Scientist's Journey Through Genesis 1–11* (RTB Press, 2014), 119–122.
22. 1 Thessalonians 1:10, 5:9; 2 Peter 2:9; Revelation 3:10.
23. John 10:17–18.
24. Genesis 15:13–16.
25. Genesis 15:16.
26. Joshua 6:17–19, 8:24–27, 10:34–40, 11:10–15.
27. Ferdinand de Saulcy, *Voyage Autour de la Mer Morte et dans les Terres Bibliques: Tome II* (Gide et J. Buadry, 1853), 20–28, archive.org/details/voyageautourdela02sauliala/page/20/mode/2up?q=Sodome; Flavius Josephus, "The Destruction of the Sodomites by the Assyrian War," in *Antiquities of the Jews*, book I, chap, 9, English translation available online at gutenberg.org/files/2848/2848-h/2848-h.htm#link2HCH0009; Chris McKinny and Kyle Keimer, hosts, *On Script Biblical World*, podcast, episode "Sodom and Gomorrah and the Cities on the Plain, Part 2," May 4, 2022, onscript.study/sodom-and-gomorrah-and-the-cities-on-the-plain-pt-2-chris-mckinny-and-kyle-keimer/.

Chapter 11: Dating Noah's Flood

1. John C. Whitcomb and Henry M. Morris, *The Genesis Flood: The Biblical Record and Its Scientific Implications* (Presbyterian & Reformed Publishing, 1961, 23rd printing 1979), 474–489.
2. Whitcomb and Morris, *The Genesis Flood*, 483–489.
3. David Wright, "Timeline for the Flood," *Answers in Genesis: Feedback* (March 9, 2012), answersingenesis.org/bible-timeline/timeline-for-the-flood/; Bodie Hodge, "Biblical Overview of the Flood Timeline," *Answers in Genesis*, August 23, 2010, answersingenesis.org/bible-timeline/biblical-overview-of-the-flood-timeline/.
4. Jonathan Sarfati, *Refuting Compromise: A Biblical and Scientific Refutation of "Progressive Creationism" (Billions of Years), as Popularized by Astronomer Hugh Ross*, 2nd ed. (Creation Book Publishers, 2011), 285.
5. Gerald E. Aardsma, *Noah's Flood Happened 3520 BC* (Aardsma Research and Publishing, 2015).
6. R. Laird Harris, Gleason L. Archer Jr., and Bruce K. Waltke, *Theological Wordbook of the Old Testament*, vol. 1 (Moody Press, 1980), 5–6.
7. Harris, Archer, and Waltke, *Theological Wordbook of the Old Testament*, vol. 1, 5–6.
8. Harris, Archer, and Waltke, 113.
9. Daniel 5:9–11, 18–22.
10. Luke 3:35–36.
11. Travis R. Freeman, "Do the Genesis 5 and 11 Genealogies Contain Gaps?," in *Coming to Grips with Genesis: Biblical Authority and the Age of the Earth*, ed. Terry Mortenson and Thane H. Ury (Master Books, 2008), 308–313; Sarfati, *Refuting Compromise*, 291–292.
12. Martin Jakobsson et al., "Post-Glacial Flooding of the Bering Land Bridge Dated to 11 cal ka BP Based on New Geophysical and Sediment Records," *Climate of the Past* 13, no. 8 (August 2017): 991–1005, doi:10.5194/cp-13-991-2017; Matthew J. Wooller et al., "A New Terrestrial Palaeoenvironmental Record from the Bering Land Bridge and Context for Human Dispersal," *Royal Society Open Science* 5, no. 6 (June 2018): id. 180145, doi:10.1098/rsos.180145; Scott A. Elias et al., "Life and Times of the Bering Land Bridge," *Nature* 382 (July 4, 1996): 61–63, doi:10.1038/382060a0; Renée Hetherington et al., "Paleogeography, Glacially Induced Crustal Displacement, and Late Quaternary Coastlines on the Continental Shelf of British Columbia, Canada," *Quaternary Science Reviews* 23, nos. 3–4 (February 2004): 295–318, doi:10.1016/j.quascirev.2003.04.001; Heiner Josenhans et al., "Early Humans and Rapidly Changing Holocene Sea Levels in the Queen Charlotte Islands–Hecate Strait, British Columbia, Canada," *Science* 277, no. 5322 (July 4, 1997): 71–74, doi:10.1126/science.277.5322.71.
13. Josenhans et al., "Early Humans and Rapidly Changing," 74.
14. Marco Baldovin et al., "Extracting Causation from Millennial-Scale Climate Fluctuations in the Last 800 kyr," *Scientific Reports* 12, no. 1 (September 12, 2022): id. 15320, doi:10.1038/s41598-022-18406-2; J. R. Petit et al., "Climate and Atmospheric History of the Past 420,000 Years from the Vostok Ice Core, Antarctica," *Nature* 399 (June 3, 1999): 429–436, doi:10.1038/20859; W. Dansgaard et al., "Evidence for General Instability of Past Climate from a 250-kyr Ice-Core Record," *Nature* 364 (July 15, 1993): 218–220, doi:10.1038/364218a0.
15. William D. Barrick, "Noah's Flood and Its Geological Implications," in *Coming to Grips with Genesis: Biblical Authority and the Age of the Earth*, ed. Terry Mortenson and Thane H. Ury (Master Books, 2008), 278.
16. Mohamed E. Hereher, "Assessment of Climate Change Impacts on Sea Surface

Temperatures and Sea Level Rise—The Arabian Gulf," *Climate* 8, no. 4 (March 30, 2020): id. 50, doi:10.3390/cli8040050.

17. Kurt Lambeck and John Chappell, "Sea Level Change Through the Last Glacial Cycle," *Science* 292, no. 5517 (April 27, 2001): 679–686, doi.10.1126/science.1059549.
18. Tamara Pico et al., "Global Ice Volume During MIS 3 Inferred from a Sea-Level Analysis of Sedimentary Core Records in the Yellow River Delta," *Quaternary Science Reviews* 152 (November 15, 2016): 72–79, doi:10.1016/j.quascirev.2016.09.012.
19. Lambeck and Chappell, "Sea Level Change," 679.
20. Jeffrey I. Rose, "New Light on Human Prehistory in the Arabo-Persian Gulf Oasis," *Current Anthropology* 51, no. 6 (December 2010): 849–883, doi:10.1086/657397.
21. Jeanna Bryner, "Lost Civilization May Have Existed Beneath the Persian Gulf," *LiveScience* (December 9, 2010), accessed January 10, 2024, livescience.com/10340-lost-civilization-existed-beneath-persian-gulf.html.
22. Hans-Peter Uerpmann and Margarethe Uerpmann, *Stone Age Sites and Their Natural Environment: The Capital Area of Northern Oman, Part 3* (Dr. Ludwig Reichert Verlag, 2004); Hans-Peter Uerpmann, Margarethe Uerpmann, and Sabah Abboud Jasim, eds., *Funeral Monuments and Human Remains From Jebel al-Buhais, The Archaeology of Jebel al-Buhais, Sharjah, United Arab Emirates,* vol. 1 (Kerns, 2006).
23. Simon J. Armitage et al., "The Southern Route 'Out of Africa': Evidence for an Early Expansion of Modern Humans into Arabia," *Science* 331, no. 6016 (January 28, 2011): 453–456, doi:10.1126/science.1199113.
24. Simon J. Greenhill, Paul Heggarty, and Russell D. Gray, "Bayesian Phylolinguistics," in *The Handbook of Historical Linguistics,* vol. 2, ed. Richard D. Janda, Brian D. Joseph, and Barbara S. Vance (Wiley-Blackwell, 2020), 238; also available at pure.mpg.de/rest/items/item_3251383/component/file_3284893/content, page 11, accessed February 21, 2025.
25. Paul Heggarty et al., "Language Trees with Sampled Ancestors Support a Hybrid Model for the Origin of Indo-European Languages," *Science* 381, no. 6656 (July 28, 2023): 415–426, doi:10.1126/science.abg0818; Remco Bouckaert et al., "Mapping the Origins and Expansion of the Indo-European Language Family," *Science* 337, no. 6097 (August 24, 2012): 957–960, doi:10.1126/science.1219669; Russell D. Gray and Quentin D. Atkinson, "Language-Tree Divergence Times Support the Anatolian Theory of Indo-European Origin," *Nature* 426, no. 6965 (November 27, 2003): 435–439, doi:10.1038/nature02029.
26. Martine Robbeets et al., "Triangulation Supports Agricultural Spread of the Transeurasian Languages," *Nature* 599, no. 7886 (November 10, 2021): 616–621, doi:10.1038/s41586-021-04108-8.
27. Greenhill, Heggarty, and Gray, "Bayesian Phylolinguistics," 240, p. 13 in the online pdf.
28. Tom Higham et al., "The Earliest Evidence for Anatomically Modern Humans in Northwestern Europe," *Nature* 479, no. 7374 (November 2, 2022): 521–524, doi:10.1038/natue10484; Sarah Pederzani et al., "Stable Isotopes Show *Homo sapiens* Dispersed into Cold Steppes ~45,000 Years Ago at Ilsenhöhle in Ranis, Germany," *Nature Ecology & Evolution* 8 (January 31, 2024): 578–588, doi:10.1038/s41559-023-02318-z; Geoff M. Smith et al., "The Ecology, Subsistence and Diet of ~45,000-Year-Old *Homo sapiens* at Ilsenhöhle in Ranis, Germany," *Nature Ecology and Evolution* 8, no. 3 (March 2024), doi:10.1038/s41559-023-02303-6; Dorothea Mylopotamitaki et al., "*Homo sapiens* Reached the Higher Latitudes of Europe by 45,000 Years Ago," *Nature* 626 (January 31, 2024): 341–346, doi:10.1038/s41586-023-06923-7; Stefano Benazzi et al., "Early Dispersal of Modern Humans in Europe and Implications for Neanderthal Behaviour," *Nature* 479 (November 2, 2011): 525–528,

doi:10.1038/nature10617; M. A. Jobling, M. E. Hurles, and C. Tyler-Smith, *Human Evolutionary Genetics: Origins, Peoples and Disease* (Garland Science, 2004), 284–285.

29. Sahra Talamo et al., "A 41,500 Year-Old Decorated Ivory Pendant from Stajnia Cave (Poland)," *Scientific Reports* 11 (November 25, 2021): id. 22078, doi:10.1038/s41598-021-01221-6; Armando Falcucci, Nicholas J. Conard, and Marco Peresani, "A Critical Assessment of the Protoaurignacian Lithic Technology at Fumane Cave and Its Implications for the Definition of the Earliest Aurignacian," *PLoS One* 12, no. 12 (December 7, 2017): e0189241, doi:10.1371/journal.pone.0189241; Jonathan A. Haws et al., "The Early Aurignacian Dispersal of Modern Humans into Westernmost Eurasia," *Proceedings of the National Academy of Sciences, USA* 117, no. 41 (September 28, 2020): 25414–25422, doi:10.1073/pnas.2016062117; Benazzi et al., "Early Dispersal of Modern Humans," 525–528; Jobling, Hurles, and Tyler-Smith, *Human Evolutionary Genetics*, 284–285.
30. Marta Mariotti Lippi et al., "Multistep Food Plant Processing at Grotta Paglicci (Southern Italy) Around 32,600 Cal B.P.," *Proceedings of the National Academy of Sciences, USA* 112, no. 39 (September 8, 2015): 12075–12080, doi:10.1073/pnas.1505213112.
31. Ceren Kabukcu et al., "Cooking in Caves: Palaeolithic Carbonised Plant Food Remains from Franchthi and Shanidar," *Antiquity* 97, no. 391 (February 2023): 12–28, doi:10.15184/aqy.2022.143.
32. Rasmus Nielsen et al., "Tracing the Peopling of the World Through Genomics," *Nature* 541 (January 18, 2017): 302–310, doi:10.1038/nature21347; Iosif Lazaridis, "The Evolutionary History of Human Populations in Europe," *Current Opinions in Genetics & Development* 53 (December 2018): 21–27, doi:10.1016/j.gde.2018.06.007; Serena Aneli et al., "Through 40,000 Years of Human Presence in Southern Europe: The Italian Case Study," *Human Genetics* 140, no. 10 (October 2021): 1417–1431, doi:10.1007/s00439-021-02328-6.
33. Higham et al., "The Earliest Evidence for Anatomically Modern Humans"; Pederzani et al., "Stable Isotopes Show *Homo sapiens* Dispersed"; Smith et al., "The Ecology, Subsistence and Diet"; Mylopotamitaki et al., "*Homo sapiens* Reached the Higher Latitudes"; Benazzi et al., "Early Dispersal of Modern Humans in Europe."
34. Qiaomei Fu et al., "Genome Sequence of a 45,000-Year-Old Modern Human from Western Siberia," *Nature* 514 (October 23, 2014): 445–449, doi:10.1038/nature13810; Solange Rigaud et al., "Symbolic Innovation at the Onset of the Upper Paleolithic in Eurasia Shown by the Personal Ornaments from Tolbor-21 (Mongolia)," *Scientific Reports* 13 (June 12, 2023): id. 9545, doi:10.1038/s41598-023-36140-1; Melinda A. Yang et al., "40,000-Year-Old Individual from Asia Provides Insight into Early Population Structure in Eurasia," *Current Biology* 27, no. 20 (October 23, 2017): 3202–3208.e9, doi:10.1016/j.cub.2017.09.030; Fabrice Demeter et al., "Anatomically Modern Human in Southeast Asia (Laos) by 46 ka," *Proceedings of the National Academy of Sciences, USA* 109, no. 36 (August 20, 2012): 14375–14380, doi:10.1073/pnas.1208104109; Graeme Barker et al., "The 'Human Revolution' in Lowland Tropical Southeast Asia: The Antiquity and Behavior of Anatomically Modern Humans at Niah Cave (Sarawak, Borneo)," *Journal of Human Evolution* 52, no. 3 (March 2007): 243–261, doi:10.1016/j.jhevol.2006.08.011; Maximilian Larena et al., "Multiple Migrations to the Philippines During the Last 50,000 Years," *Proceedings of the National Academy of Sciences, USA* 118, no. 13 (March 22, 2021): id. e2026132118, doi:10.1073/pnas.2026132118; Raj Pradip Chakraborty, "Human Dispersals in South Asia: A Genetic Odyssey," *Conference: Conflicts, Issues, and Challenges in South Asia with Special Reference to North Eastern Region of India* (October 2017), researchgate.net/publication/328637784_Human_dispersals_in_South_AsiaA_genetic_odyssey.
35. J. F. O'Connell and J. Allen, "The Process, Biotic Impact, and Global Implications of the

Human Colonization of Sahul About 47,000 Years Ago," *Journal of Archaeological Science* 56 (April 2015): 73–84, doi:10.1016/j.jas.2015.02.020; Anna-Sapfo Malaspinas et al., "A Genomic History of Aboriginal Australia," *Nature* 538 (October 13, 2016): 207–214, doi:10.1038/nature18299.

36. Alia J. Lesnek et al., "Deglaciation of the Pacific Coastal Corridor Directly Preceded the Human Colonization of the Americas," *Science Advances* 4, no. 5 (May 30, 2018): id. eaar5040, doi:10.1126/sciadv.aar5040; Michael R. Waters, "Late Pleistocene Exploration and Settlement of the Americas by Modern Humans," *Science* 365, no. 6449 (July 12, 2019): id. Eaat5447, doi:10.1126/science.aat5447; Bastien Llamas et al., "Ancient Mitochondrial DNA Provides High-Resolution Time Scale of the Peopling of the Americas," *Science Advances* 2, no. 4 (April 1, 2016): id. e1501385, doi:10.1126/sciadv.1501385; Todd A. Surovell et al., "Late Date of Human Arrival to North America: Continental Scale Differences in Stratigraphic Integrity of Pre-13,000 BP Archaeological Sites," *PLoS One* 17, no. 4 (April 20, 2022): id. e0264092, doi:10.1371/journal.pone.0264092; Ted Goebel, Michael R. Waters, and Dennis H. O'Rourke, "The Late Pleistocene Dispersal of Modern Humans in the Americas," *Science* 319, no. 5869 (March 14, 2008): 1497–1502, doi:10.1126/science.1153569.
37. Hie Lim Kim et al., "Prehistoric Human Migration Between Sundaland and South Asia Was Driven by Sea-Level Rise," *Communications Biology* 6 (February 4, 2023): id. 150, doi:10.1038/s42003-023-04510-0; Michael I. Bird, David Taylor, and Chris Hunt, "Palaeoenvironments of Insular Southeast Asia During the Last Glacial Period: A Savanna Corridor in Sundaland?," *Quaternary Science Reviews* 24, nos. 20–21 (November 2005): 2228–2242, doi:10.1016/j.quascirev.2005.04.004; Vincent Macaulay et al., "Single, Rapid Coastal Settlement of Asia Revealed by Analysis of Complete Mitochondrial Genomes," *Science* 308, no. 5724 (May 13, 2005): 1034–1036, doi:10.1126/science.1109792; Lesnek et al., "Deglaciation of the Pacific Coastal Corridor"; E. James Dixon, "Late Pleistocene Colonization of North America from Northeast Asia: New Insights from Large-Scale Paleogeographic Reconstructions," in *Mobility and Ancient Society of Asia and the Americas*, ed. Michael David Franchetti and Robert N. Spengler III (Springer, 2015), 169–184, doi:10.1007/978-3-319-15138-0_12; John S. Compton, "Pleistocene Sea-Level Fluctuations and Human Evolution on the Southern Coastal Plain of South Africa," *Quaternary Science Reviews* 30, nos. 5–6 (March 2011): 506–527, doi:10.1016/j.quascirev.2010.12.012.
38. Ash Parton et al., "Alluvial Fan Records from Southeast Arabia Reveal Multiple Windows for Human Dispersal," *Geology* 43, no. 4 (April 1, 2015): 295–2908, doi:10.1130/G36401.1.
39. Huw S. Groucutt et al., "Human Occupation of the Arabian Empty Quarter During MIS 5: Evidence from Mundafan Al-Buhayrah, Saudi Arabia," *Quaternary Science Reviews* 119 (July 1, 2015): 116–135, doi:10.1016/j.quascirev.2015.04.020; M. Sultan et al., "Geochemical, Isotopic, and Remote Sensing Constraints on the Origin and Evolution of the Rub Al Khali Aquifer System, Arabian Peninsula," *Journal of Hydrology* 356, nos. 1–2 (July 1, 2008): 70–83, doi:10.1016/j.jhydrol.2008.04.001; Ian D. Clark and Jean-Charles Fontes, "Paleoclimatic Reconstruction in Northern Oman Based on Carbonates from Hyperalkaline Groundwaters," *Quaternary Research* 33, no. 3 (May 1990): 320–336, doi:10.1016/0033-5894(90)90059-T.
40. Paul S. Breeze et al., "Palaeogydrological Corridors for Hominin Dispersals in the Middle East ~250–70,000 Years Ago," *Quaternary Science Reviews* 144 (July 15, 2016): 155–185, doi:10.1016/j.quascirev.2016.05.012; Matthew Stewart et al., "Human Footprints Provide Snapshot of Last Interglacial Ecology in the Arabian Interior," *Science Advances* 6, no. 38 (September 18, 2020): id. eaba8940, doi:10.1126/sciadv.aba8940.
41. Stewart et al., "Human Footprints Provide Snapshot."

42. Ignacio A. Lazagabaster et al., "Rare Crested Rat Subfossils Unveil Afro-Eurasian Ecological Corridors Synchronous with Early Human Dispersals," *Proceedings of the National Academy of Sciences, USA* 118, no. 31 (July 26, 2021): e2105719118, doi:10.1073/pnas.2105719118.
43. B. C. C. Hume et al., "*Symbiodinium thermophilum* sp. nov., a Thermotolerant Symbiotic Alga Prevalent in Corals of the World's Hottest Sea, the Persian/Arabian Gulf," *Scientific Reports* 5 (February 27, 2015): id. 8562, doi:10.1038/srep08562.
44. B. Hume et al., "Corals from the Persian/Arabian Gulf as Models for Thermotolerant Reef-Builders: Prevalence of Clade C3 *Symbiodinium*, Host Fluorescence and *Ex Situ* Temperature Tolerance," *Marine Pollution Bulletin* 72, no. 2 (July 30, 2013): 313–322, doi:10.1016/j.marpolbul.2012.11.032; Charles Sheppard et al., "The Gulf: A Young Sea in Decline," *Marine Pollution Bulletin* 60, no. 1 (January 2010): 13–38, doi:10.1016/j.marpolbul.2009.10.017.
45. R. L. K. Fullagar, D. M. Price, and L. M. Head, "Early Human Occupation of Northern Australia: Archaeology and Thermoluminescence Dating of Jinmium Rock-Shelter, Northern Territory," *Antiquity* 70, no. 270 (December 1996): 751–773, doi:10.1017/S0003598X00084040.
46. Richard Roberts et al., "Optical and Radiocarbon Dating at Jinmium Rock Shelter in Northern Australia," *Nature* 393 (May 28, 1998): 358–362, doi:10.1038/30718.

Chapter 12: Biblical Clues to the Flood's Extent

1. Simon G. Southerton, *Losing a Lost Tribe: Native Americans, DNA, and the Mormon Church* (Signature Books, 2004).
2. Romans 1:8.
3. Colossians 1:6.
4. Jonathan Sarfati, *Refuting Compromise: A Biblical and Scientific Refutation of "Progressive Creationism" (Billions of Years), As Popularized by Astronomer Hugh Ross*, 2nd ed. (Creation Book Publishers, 2011), 249; William D. Barrick, "Noah's Flood and Its Geological Implications," in *Coming to Grips with Genesis: Biblical Authority and the Age of the Earth*, ed. Terry Mortenson and Thane H. Ury (Master Books, 2008), 255–257, 264; Ken Ham and Bodie Hodge, *A Flood of Evidence: 40 Reasons Noah and the Ark Still Matter* (Master Books, 2016), 92–96.
5. R. Laird Harris, Gleason L. Archer Jr., and Bruce K. Waltke, *Theological Wordbook of the Old Testament*, vol. 1 (Moody Press, 1980), 135–136.
6. Harris, Archer, and Waltke, *Theological Wordbook*, vol. 1, 92–93.
7. Harris, Archer, and Waltke, 281.
8. R. Laird Harris, Gleason L. Archer Jr., and Bruce K. Waltke, *Theological Wordbook of the Old Testament*, vol. 2, (Moody Press, 1980), 587–591.
9. Harris, Archer, and Waltke, *Theological Wordbook*, vol. 2, 654–655.
10. Harris, Archer, and Waltke, 850–851.
11. Harris, Archer, and Waltke, 775.
12. Hugh Ross, *Hidden Treasures in the Book of Job: How the Oldest Book in the Bible Answers Today's Scientific Questions* (Baker Books, 2011), 131–173.
13. Harris, Archer, and Waltke, *Theological Wordbook*, vol. 1, 449.
14. *EF English Live*, "How Many Words Are in the English Language?," *Language Lab*, accessed September 19, 2023, englishlive.ef.com/blog/language-lab/many-words-english-language/. Biochemical terms and the names of species of life add more than two million words.
15. Harris, Archer, and Waltke, *Theological Wordbook*, vol. 1, 224–225.
16. Harris, Archer, and Waltke, *Theological Wordbook*, 146; Gesenius, *Gesenius' Hebrew-Chaldee Lexicon to the Old Testament,* trans. Samuel Prideaux Tregelles (Baker Book House,

1979), 153.

17. Leonard J. Coppes in Harris, Archer, and Waltke, *Theological Wordbook*, vol. 1, 10.
18. Harris, Archer, and Waltke, *Theological Wordbook*, vol. 1, 319.
19. Terry Mortenson, "Psalm 104:6–9—the Flood or Day Three of Creation Week?," *Answers in Genesis,* July 28, 2009, accessed January 2, 2025, answersingenesis.org/the-flood/psalm-104-flood-or-day-three-of-creation-week/.
20. The six creation psalms are Psalms 8, 19, 65, 104, 147, 148.
21. Kenneth S. Wuest, *Wuest's Word Studies from the Greek New Testament: In These Last Days*, vol. 2 (Eerdmans, 1973), 51.
22. Wuest, *Wuest's Word Studies*, 51.
23. John C. Whitcomb and Henry M. Morris, *The Genesis Flood: The Biblical Record and Its Scientific Implications* (Presbyterian & Reformed Publishing, 1961), 44, 451–453; Henry M. Morris, *Biblical Creationism: What Each Book of the Bible Teaches About Creation and the Flood* (Master Books, 2000), 187–189; Simon Turpin, "How the Apostle Peter Relates to the Age of the Earth Debate: Analyzing 2 Peter 3:1–7," *Answers in Genesis*, June 1, 2016, accessed September 19, 2023, answersingenesis.org/days-of-creation/apostle-peter-relates-age-earth-debate/; Keaton Halley and Gary Bates, "Faltering on the Flood: Evading the Bible's Clear Meaning Is Disastrous," *CMI Newsletter* (July 2016), published online May 18, 2017, accessed September 19, 2023, creation.com/local-flood.
24. Luke 17:29.
25. Genesis 5:3–27.
26. Hugh Ross, *Navigating Genesis: A Scientist's Journey Through Genesis 1–11* (RTB Press, 2014), 119–122.
27. Genesis 11:4.
28. Harris, Archer, and Waltke, *Theological Wordbook*, vol. 1, 309; Harris, Archer, and Waltke, *Theological Wordbook*, vol. 2, 800, 809, 923; Gesenius, *Gesenius' Hebrew-Chaldee Lexicon*, 295, 733, 807–809, 821.
29. Hebrews 11:7.
30. 2 Peter 2:5.
31. Larry Vardiman, Andrew A. Snelling, and Eugene F. Chaffin, eds., *Radioisotopes and the Age of the Earth: A Young-Earth Creationist Research Initiative*, vol. 1 (Institute for Creation Research; Creation Research Society, 2000), 307, accessed January 2, 2025, icr.org/i/pdf/research/rate-all.pdf.
32. Hugh Ross, *A Matter of Days: Resolving a Creation Controversy*, 2nd ed. (RTB Press, 2015), 139.

Chapter 13: Scientific Clues to the Flood's Extent

1. Christian Reformed Church, *Ecumenical Creeds and Reformed Confessions* (CRC Publications, 1988), 79.
2. Hugh Ross, *Rescuing Inerrancy: A Scientific Defense* (RTB Press, 2023).
3. Jerry A. Coyne, *Faith vs. Fact: Why Science and Religion Are Incompatible* (Viking, 2015), 124.
4. Gregg Davidson, "Science vs. Flood Geology: Not Just a Difference in Worldview," *BioLogos: The Flood* (July 31, 2018), accessed October 17, 2023, biologos.org/articles/science-vs-flood-geology-not-just-a-difference-in-worldview.
5. Davidson, "Science vs. Flood Geology."
6. Genesis 7:11, 8:5–6, 13–14.

7. Mohammad Hadi Bordbar et al., "The Persian Gulf and Oman Sea: Climate Variability and Trends Inferred from Satellite Observations," *Estuarine, Coastal and Shelf Science* 296 (January 2024): id. 108588, doi:10.1016/j.ecsss.2023.108588.
8. David MacDonald, "The Flood: Mesopotamian Archaeological Evidence," *Creation/Evolution Journal* 8, no. 2 (Spring 1988): 14–20, accessed October 19, 2023, ncse.ngo/flood-mesopotamian-archaeological-evidence; T. C. Mitchell, "Geology and the Flood" in *New Bible Dictionary*, 2nd ed., ed. J. D. Douglas et al. (InterVarsity Press, 1982), 382–383.
9. Ken Ham and Bodie Hodge, *A Flood of Evidence: 40 Reasons Noah and the Ark Still Matter* (Master Books, 2016); Andrew A. Snelling, "Geologic Evidences for the Genesis Flood, Part I: An Overview," *Answers Magazine*, October–December 2007, accessed February 21, 2025, answersingenesis.org/the-flood/geologic-evidences-for-the-genesis-flood/; D. Russell Humphreys, "Reversals of the Earth's Magnetic Field During the Genesis Flood," in *Proceedings of the First International Conference on Creationism*, vol. 1, article 52, ed. Robert E. Walsh, Christopher L. Brooks, and Richard S. Crowell (Creation Science Fellowship, 1986), 113–126.
10. Mathew Domeier, Rob Van der Voo, and Trond H. Torsvik, "Paleomagnetism and Pangea: The Road to Reconciliation," *Tectonophysics* 514–517 (January 5, 2012): 14–43, doi:10.1016/j.tecto.2011.10.021.
11. Masaki Yoshida and Yozo Hamano, "Pangea Breakup and Northward Drift of the Indian Subcontinent Reproduced by a Numerical Model of Mantle Convection," *Scientific Reports* 5 (February 12, 2015): id. 8407, doi:10.1038/srep08407.
12. Prakash Kumar et al., "The Rapid Drift of the Indian Tectonic Plate," *Nature* 449 (October 18, 2007): 894–897, doi:10.1038/nature06214.
13. Carol Hill ed., et al., *The Grand Canyon: Monument to an Ancient Earth* (Kregel Publications, 2016).
14. Robert M. Carter and Paul Gammon, "New Zealand Maritime Glaciation: Millennial-Scale Southern Climate Change Since 3.9 Ma," *Science* 304, no. 5677 (June 11, 2004): 1659–1662, doi:10.1126/science.1093726.
15. S. Robertson, "BGS Rock Classification Scheme, Volume 2, Classification of Metamorphic Rocks," *British Geological Survey, Research Report Number RR 99-02* (British Geological Survey, Keyworth: 1999), accessed October 31, 2023, nora.nerc.ac.uk/id/eprint/3226/1/RR99002.pdf.
16. Ham and Hodge, *A Flood of Evidence*, 164; Antonio Snider-Pellegrini, *Le Création et ses Mystères Dévoilés* (Franck and Dentu, 1859).
17. Ham and Hodge, *A Flood of Evidence*, 164; Andrew A. Snelling, "Noah's Lost World," *Answers Magazine*, April–June 2014, accessed November 8, 2023, answersingenesis.org/geology/plate-tectonics/noahs-lost-world/.
18. Zhen-Jie Zhang et al., "Lithospheric Thickness Records Tectonic Evolution by Controlling Metamorphic Conditions," *Science Advances* 9, no. 50 (December 15, 2023): id. eadi2134, doi:10.1126/sciadv.adi2134.
19. John R. Baumgardner, "Catastrophic Plate Tectonics: The Physics Behind the Genesis Flood," *Proceedings of the Fifth International Conference on Creationism*, ed. R. I. Ivey, Jr. (Creation Science Fellowship, 2003), vol. 5, article 13, 113–126, digitalcommons.cedarville.edu/icc_proceedings/vol5/iss1/13; John Baumgardner, "Catastrophic Plate Tectonics: The Geophysical Context of the Genesis Flood," *Journal of Creation: Technical Journal* 16, no. 1 (April 2002): 58–63, creation.com/catastrophic-plate-tectonics-the-geophysical-context-of-the-genesis-flood; Steven A. Austin et al., "Catastrophic Plate Tectonics: A Global Flood

Model of Earth History," *Proceedings of the Third International Conference on Creationism*, held July 18–23, 1994, ed. Robert E. Walsh (Creation Science Fellowship, 1994), 609–621, digitalcommons.cedarville.edu/icc_proceedings/vol3/iss1/56; William D. Barrick, "Noah's Flood and Its Geological Implications," in *Coming to Grips with Genesis: Biblical Authority and the Age of the Earth*, ed. Terry Mortenson and Thane H. Ury (Master Books, 2008), 275–281.

20. Baumgardner, "Catastrophic Plate Tectonics: The Physics Behind the Genesis Flood," 113–126; Baumgardner, "Catastrophic Plate Tectonics: The Geophysical Context of the Genesis Flood," 58–63; Austin et al., "Catastrophic Plate Tectonics: A Global Flood Model of Earth History," 609–621; Barrick, "Noah's Flood and Its Geological Implications," 275–281.
21. Walter T. Brown, *In the Beginning: Compelling Evidence for Creation and the Flood*, 5th ed. (Center for Scientific Creation, 1989), 58–84; Jonathan Sarfati, *Refuting Compromise: A Biblical and Scientific Refutation of "Progressive Creationism" (Billions of Years), as Popularized by Astronomer Hugh Ross*, 2nd ed. (Creation Book Publishers, 2011), 273–274, 279–280, 368–370.
22. Timothy H. Heaton, "Recent Developments in Young-Earth Creationist Geology," *Science & Education* 18, no. 10 (October 2009): 1341–1358, doi:10.1007/s11191-008-9162-6.
23. European Space Agency, "Earth Movements from Japan Earthquake Seen from Space," *ESA/Applications/Observing the Earth/Envisat* (March 30, 2011), accessed November 5, 2023, esa.int/Applications/Observing_the_Earth/Envisat/Earth_movements_from_Japan_earthquake_seen_from_space.
24. K. E. Bullen, *An Introduction to the Theory of Seismology* (Cambridge University Press, 1963), 259–262; Hiroo Kanamori, "Shaking Without Quaking," *Science* 279, no. 5359 (March 27, 1998): 2063–2064, doi:10.1126/science.279.5359.2063.
25. Larry Vardiman, Andrew A. Snelling, and Eugene F. Chaffin, eds., *Radioisotopes and the Age of the Earth: A Young-Earth Creationist Research Initiative*, vol. 1 (Institute for Creation Research; Creation Research Society, 2000); Larry Vardiman, Andrew A. Snelling, and Eugene F. Chaffin, eds., *Radioisotopes and the Age of the Earth: A Young-Earth Creationist Research Initiative*, vol. 2 (Institute for Creation Research; Creation Research Society, 2005).
26. Hugh Ross, *A Matter of Days: Resolving a Creation Controversy*, 2nd ed. (RTB Press, 2015), 163–166, 172–176.
27. Ross, *A Matter of Days*, 165.
28. Daniel M. Sigman, Mathis P. Hain, and Gerald H. Haug, "The Polar Ocean and Glacial Cycles in Atmospheric CO_2 Concentration," *Nature* 466 (July 1, 2010): 47–55, doi:101038/nature09149; EPICA community members, Laurent Augustin et al., "Eight Glacial Cycles from an Antarctic Ice Core," *Nature* 429 (June 10, 2004): 623–628, doi:10.1038/nature02599; Hideaki Motoyama, "The Second Deep Ice Coring Project at Dome Fuji, Antarctica," *Scientific Drilling* 5 (September 2007): 41–43, doi:10.5194/sd-5-41-2007.
29. North Greenland Ice Core Project members, K. K. Andersen et al., "High-Resolution Record of Northern Hemisphere Climate Extending into the Last Interglacial Period," *Nature* 431 (September 9, 2004): 147–151, doi:10.1038/nature02805.
30. Andrew A. Snelling and Mike Matthews, "When Was the Ice Age in Biblical History?," *Answers Magazine*, April 2013, 46–52, accessed November 13, 2023, answersingenesis.org/environmental-science/ice-age/when-was-the-ice-age-in-biblical-history/; Larry Vardiman, "An Analytical Young-Earth Flow Model of the Ice Sheet Formation During the 'Ice Age,'" *Proceedings of the Third International Conference on Creationism* vol. 3, no. 60, ed. Robert Walsh (Creation Science Fellowship, 1994),

561–568, accessed November 13, 2023, digitalcommons.cedarville.edu/cgi/viewcontent.cgi?article=1361&context=icc_proceedings; Jake Hebert, "Ice Cores, Seafloor Sediments, and the Age of the Earth, Part 2," *Acts & Facts, Impact* 43, no. 7 (June 30, 2014), accessed November 13, 2023, icr.org/article/8181; Ham and Hodge, *A Flood of Evidence*, 156.

31. Michael Oard, *An Ice Age Caused by the Genesis Flood* (Institute for Creation Research, 1990); Michael J. Oard, "An Ice Age Within the Biblical Time Frame," in *Proceedings of the First International Conference on Creationism II* , vol. 1, no. 48, ed. R. E. Walsh, Christopher L. Brooks, and Richard S. Crowell, (Creation Science Fellowship, 1986), 157–163; Vardiman, "An Analytical Young-Earth Flow Model," 561–568.
32. Carl Wieland, "The Lost Squadron: Deeply Buried Missing Planes Challenge 'Slow and Gradual' Preconceptions," *Creation* 19, no. 3 (June 1997): 10–14, creation.com/the-lost-squadron.
33. Carter and Gammon, "New Zealand Maritime Glaciations," 1659–1662.
34. J. S. Dean, "Dendrochronology," in *Dating and Age Determination of Biological Materials*, ed. Michael R. Zimmerman and J. Lawrence Angel (Croom Helm, 1986), 126–165.
35. M. G. L. Baillie and J. R. Pilcher, "The Belfast 'Long Chronology' Project," in *Applications of Tree-Ring Studies: Current Research in Dendrochronology and Related Subjects*, BAR International Series 333, ed. R. G. W. Ward (British Archaeological Reports, May 1987): 203–214; Bernd Becker, "An 11,000-Year German Oak and Pine Dendrochronology for Radiocarbon Calibration," *Radiocarbon* 35, no. 1 (1993): 201–213, doi:10.1017/S0033822200012898.
36. Michael R. May et al., "A Pleistocene Clone of Palmer's Oak Persisting in Southern California," *PLoS ONE* 4, no. 12 (December 23, 2009): e8346, doi:10.1371/journal.pone.0008346.
37. Larry Vardiman, "RATE in Review: Unresolved Problems," *Acts & Facts* 36, no. 12 (December 2007): 6, accessed October 27, 2023, icr.org/article/rate-review-unresolved-problems/.
38. Russell Humphreys quoted by Larry Vardiman, in Vardiman, "RATE in Review."
39. Max J. Hunter, "The Pre-Flood/Flood Boundary at the Base of the Earth's Transition Zone," *Journal of Creation* 14, no. 1 (April 2000): 60–74, creation.com/the-pre-flood-flood-boundary-at-the-base-of-the-earths-transition-zone, accessed February 21, 2025; Glenn R. Morton, "The Flood on an Expanding Earth," *Creation Research Society Quarterly* 19, no. 4 (March 1983): 219–224; also available at creationresearch.org/crsq-1983-volume-19-number-4_flood-on-an-expanding-earth, accessed February 21, 2025; Glenn R. Morton, "Mountain Synthesis on an Expanding Earth," *Creation Research Society Quarterly* 24, no. 2 (September 1987): 53–58; also available at creationresearch.org/crsq-1987-volume-24-number-2_mountain-synthesis, accessed February 21, 2025.
40. Brandon Schmandt et al., "Dehydration Melting at the Top of the Lower Mantle," *Science* 344, no. 6189 (June 13, 2014): 1265–1268, doi:10.1126/science.1253358; Takashi Nakagawa, "Numerical Modeling on Global-Scale Mantle Water Cycle and Its Impact on the Sea-Level Change," *Earth and Planetary Science Letters* 619 (October 1, 2023): id. 118312, doi:10.1016/j.epsl.2023.118312; Hongzhan Fei, "Water Content of the Dehydration Melting Layer in the Topmost Lower Mantle," *Geophysical Research Letters* 48, no. 1 (January 16, 2021): id. E2020GL090973, doi:10.1029/2020GL090973, Matthew G. Price, J. H. Davies, and James Panton, "Controls on the Deep-Water Cycle Within Three-Dimensional Mantle Convection Models," *Geochemistry, Geophysics, Geosystems* 20, no. 5 (May 2019): 2199–2212, doi:10.1029/2018GC008158.
41. Earth's magnetic field strength does not steadily decline. Rather, it weakens, builds up, weakens, and builds up in a semiperiodic sinusoidal pattern. Over the past 12,000 years, Earth's magnetic field measures to have varied between 85 and 150 percent of its present value.

42. Humphreys, "Reversals of the Earth's Magnetic Field," 113–126; D. Russell Humphreys, "The Earth's Magnetic Field Is Young," *Acts and Facts*, 22 no. 8 (August 1993): i–iv, icr.org/article/earths-magnetic-field-young/.
43. Evan J. Gowan et al., "A New Global Ice Sheet Reconstruction for the Past 80,000 Years," *Nature Communications* 12 (February 23, 2021): id. 1199, doi:10.1038/s41467-021-21469-w.
44. Carol A. Hill, "Qualitative Hydrology of Noah's Flood," *Perspectives on Science and Christian Faith* 58, no. 2 (June 2006): 123, accessed January 2, 2025, asa3.org/ASA/PSCF/2006/PSCF6-06Hill.pdf.
45. Ali Mohammadi, "Aeolian and Fluvial Processes Influence on Dust Storms of Hormuz Strait and Makran Coastal Plains (SE Iran); Insight from Geomorphic Landforms, and Sediment Texture and Mineralogy," *International Journal of Earth Sciences* 112, no. 7 (August 2023): 1973–1987, doi:10.1007/s00531-023-02335-0; S. Mostafa Mousavi, "Mapping Seismic Moment and b-Value Within the Continental-Collision Orogenic-Belt Region of the Iranian Plateau," *Journal of Geodynamics* 103 (January 2017): 26–41, doi:10.1016/j.jog.2016.12.001; Charlotte Fillon, Daniel Garcia-Castellanos, and Jaume Vergés, "Coupled Tectonic/Surface Processes Modeling of the Neogene Mesopotamian Basin Evolution," *EGU General Assembly 2014*, held April 27–May 2, 2014 in Vienna, Austria (May 2014): id. 10573, accessed January 2, 2025, meetingorganizer.copernicus.org/EGU2014/EGU2014-10573.pdf.
46. Stephen A. Swift and Amy S. Bower, "Formation and Circulation of Dense Water in the Persian/Arabian Gulf," *Journal of Geophysical Research: Oceans* 108, no. C1 (January 2003): 4-1–4-21, doi:10.1029/2002JC001360.
47. Roelof D. Schuiling et al., "The Hormuz Strait Dam Macroproject—21st Century Electricity Development Infrastructure Node (EDIN)?," *Marine Georesources & Geotechnology* 23, nos. 1–2 (January 2005): 25–37, doi:10.1080/10641190590959920.

Chapter 14: The Necessity and Feasibility of the Ark

1. Matthew 24:38–39.
2. Sir Charles Warren, *The Ancient Cubit and Our Weights and Measures* (Committee of the Palestine Exploration Fund, 1903), 4–10, 111–120, accessed August 27, 2023, archive.org/details/ancientcubitourw00warruoft/page/120/mode/2up.
3. Warren, *The Ancient Cubit*, 4–10, 111–120.
4. Warren, 4–10, 111–120.
5. Carol A. Hill, "A Time and a Place for Noah," *Perspectives on Science and Christian Faith* 53, no. 1 (March 2001): 35, accessed January 2, 2025, asa3.org/ASA/PSCF/2001/PSCF3-01Hill.html.
6. Athenaeus of Naucratis, *The Deipnosophists*, vol. 1, trans. C. D. Yonge (Henry G. Bohn, 1854), 324–325; Plutarch, *The Parallel Lives: The Life of Demetrius*, section 43, paragraph 4, in *Plutarch, Volume IX, Loeb Classical Library*, trans. Bernadotte Perrin (Harvard University Press, 1920), 109, penelope.uchicago.edu/Thayer/E/Roman/Texts/Plutarch/Lives/Demetrius*.html.
7. Eric Meier, "World's Strongest Woods," *The Wood Database* (2023), accessed October 15, 2023, wood-database.com/worlds-strongest-woods/.
8. Meier, "World's Strongest Woods."
9. Meier.
10. Meier.
11. Meier.

12. Meier.
13. Gerald A. Larue, "Flood Myths and Sunken Arks: Who Needs to Believe in Noah's Ark and Why?," *Skeptic Magazine*, 1994, accessed January 2, 2025, skeptic.com/reading_room/flood-myths-who-believes-in-noahs-ark-and-why/; Jeremy Bowen, "Did Noah Really Build an Ark?," *BBC News: One-Minute World News*, March 19, 2004, accessed May 1, 2024, news.bbc.co.uk/2/hi/uk_news/magazine/3524676.stm.
14. Riczar Fuentes and Alfred Pawlik, "Testing the Waters: Plant Working and Seafaring in Pleistocene Wallacea," *Journal of Archaeological Science: Reports* 62 (April 2025): id. 105020, doi:10.1016/j.jasrep.2025.105020; Ceri Shipton et al., "Abrupt Onset of Intensive Human Occupation 44,000 Years Ago on the Threshold of Sahul," *Nature Communications* 15, no. 1 (May 22, 2024): id. 4193, doi:10.1038/s41467-024-48395-x; Sue O'Connor et al., "Terminal Pleistocene Emergence of Maritime Interaction Networks Across Wallacea," *World Archaeology* 54, no. 2 (March 15, 2022): 244–263, doi:10.1080/00438243.2023.2172072.
15. Ceren Kabukcu et al., "Cooking in Caves: Palaeolithic Carbonised Plant Food Remains from Franchthi and Shanidar," *Antiquity* 97, no. 391 (February 2023): 12–28, doi:10.15184/aqy.2022.143.
16. Efraim Lev, Mordechai E. Kislev, and Ofer Bar-Yosef, "Mousterian Vegetal Food in Kebara Cave, Mt. Carmel," *Journal of Archaeological Science* 32, no. 3 (March 2005): 475–484, doi:10.1016/j.jas.2004.11.006.
17. Alan E. Hill, "Quantitative Hydrology of Noah's Flood," *Perspectives on Science and Christian Faith* 58, no. 2 (June 2006): 130–141, accessed January 2, 2025, asa3.org/ASA/PSCF/2006/PSCF6-06Hill2.pdf.
18. Fernand Navarra, *Noah's Ark: I Touched It*, ed. Dave Balsinger (Logos International, 1974); Violet M. Cummings, *Noah's Ark: Fact or Fable?* (Creation-Science Research Center, 1972); Science News, "In the Wake of the Ark," *Science News* 97, no. 24 (June 13, 1970): 574, sciencenews.org/archive/wake-ark; Kelly L. Segraves, *Search for Noah's Ark* (Beta Books, 1975); *In Search of Noah's Ark*, directed by James L. Conway (Sunn Classic Pictures, 1976); Andrew A. Snelling, "Is the Wood Recently Found on Mt. Ararat from the Ark?," *Answers in Depth* 6 (November 9, 2011), accessed November 14, 2023, answersingenesis.org/noahs-ark/noahs-ark-found/is-the-wood-recently-found-on-mt-ararat-from-the-ark/.
19. Ron Wyatt, *Discovered: Noah's Ark* (World Bible Society, 1989); Allen Roberts, *Arksearch: The Terrifying Quest* (Monarch Publications, 1994); David Fasold, *The Ark of Noah* (Wynwood, 1988); Tim Newcomb, "Archaeologists Think They Might Have Found the Real Noah's Ark," *Popular Mechanics*, November 1, 2023, accessed November 14, 2023, popularmechanics.com/science/archaeology/a45700571/was-noahs-ark-found/.

Chapter 15: "Fantastic" or Factual

1. William Lane Craig, *In Quest of the Historical Adam: A Biblical and Scientific Exploration* (Eerdmans, 2021), 104–131.
2. Craig, *In Quest of the Historical Adam*, 104–105.
3. Craig, 131.
4. Craig, 111–129.
5. Craig, 111.
6. Christopher E. Gordon et al., "Mesopredator Suppression by an Apex Predator Alleviates the Risk of Predation Perceived by Small Prey," *Proceedings of the Royal Society B* 282, no. 1802 (March 7, 2015): id. 20142870, doi:10.1098/rspb.2014.2870; Emma Marris, "Rethinking Predators: Legend of the Wolf," *Nature* 507 (March 13, 2014): 158–160, doi:10.1038/507158a;

Adam Pérou Hermans et al., "Wolf Reintroduction: Ecological Management and the Substitution Problem," *Ecological Restoration* 32, no. 3 (September 1, 2014): 221–228, doi:10.3368/er.32.3.221; William J. Ripple and Robert L. Beschta, "Wolves and the Ecology of Fear: Can Predation Risk Structure Ecosystems?," *BioScience* 54, no. 8 (August 2004): 755–766, doi:10.1641/0006-3568(2004)054[0755:WATEOF]2.0.CO;2; Hugh Ross, "More Reasons to Thank God for Apex Predators," *Today's New Reason to Believe* (blog), Reasons to Believe, August 15, 2016, reasons.org/explore/blogs/todays-new-reason-to-believe/more-reasons-to-thank-god-for-apex-predators.

7. R. C. Sproul, ed., *The Reformation Study Bible* (Ligonier Ministries, 2005), 8.
8. Craig, *In Quest of the Historical Adam*, 111.
9. R. Laird Harris, Gleason L. Archer Jr., and Bruce K. Waltke, *Theological Wordbook of the Old Testament*, vol. 1 (Moody Press, 1980), 918a.
10. R Laird Harris, Gleason L. Archer Jr., and Bruce K. Waltke, *Theological Wordbook of the Old Testament*, vol. 2 (Moody Press, 1980), 1707a.
11. Susan Solomon, "Risks to the Stratospheric Ozone Shield in the Anthropocene," *Ambio* 50, no. 1 (January 2021): 44–48, doi:10.1007/s13280-020-01431-8; Harry Slaper et al., "Estimates of Ozone Depletion and Skin Cancer Incidence to Examine the Vienna Convention Achievements," *Nature* 384 (November 21, 1996): 256–258, doi:10.1038/384256a0; Harold S. Johnston, "Analysis of the Independent Variables in the Perturbation of Stratospheric Ozone by Nitrogen Fertilizers," *Journal of Geophysical Research* 82, no. 12 (April 20, 1977): 1767–1772, doi:10.1029/JC082i012p01767; Paul J. Crutzen and Dieter H. Ehhalt, "Effects of Nitrogen Fertilizers and Combustion on the Stratospheric Ozone Layer," *Ambio* 6, nos. 2–3 (January 1, 1977): 112–117, jstor.org/stable/4312257.
12. Eileen Abt and Lauren Posnick Robin, "Perspective on Cadmium and Lead in Cocoa and Chocolate," *Journal of Agricultural and Food Chemistry* 68, no. 46 (November 18, 2020): 13008–13015, doi:10.1021/acs.jafc.9b08295; Margaret R. Karagas et al., "Rice Intake and Emerging Concerns on Arsenic in Rice: A Review of the Human Evidence and Methodologic Challenges," *Current Environmental Health Reports* 6, no. 4 (December 2019): 361–372, doi:10.1007/s40572-019-00249-1.
13. A. D. Erlykin and A. W. Wolfendale, "A Single Source of Cosmic Rays in the Range 10^{15}–10^{16} eV," *Journal of Physics G: Nuclear and Particle Physics* 23, no. 8 (August 1997): 979–989, doi:10.1088/0954-3899/23/8/012; A. D. Erlykin and A. W. Wolfendale, "High Energy Cosmic Ray Mass Spectroscopy. II. Masses in the Range 10^{14}–10^{17} eV," *Astroparticle Physics* 7, no. 3 (August 1997): 203–211, doi:10.1016/S0927-6505(97)00020-0; A. D. Erlykin and A. W. Wolfendale, "Structure in the Cosmic Ray Spectrum: An Update," *Journal of Physics G: Nuclear and Particle Physics* 27, no. 5 (May 2001): 1005–1030, doi:10.1088/0954-3899/27/5/305; A. D. Erlykin and A. W. Wolfendale, "Spectral Features and Masses in the PeV Region." *Nuclear Physics B—Proceedings Supplements* 122 (July 2003): 209–212, doi:10.1016/S0920-5632(03)80380-8.
14. A. D. Erlykin and A. W. Wolfendale, "Further Evidence Favouring the Single Source Model for Cosmic Rays," *Astroparticle Physics* 23, no. 1 (February 2005): 1–9, doi:10.1016/j.astropartphys.2004.09.009.
15. M. Amenomori et al., The Tibet ASg Collaboration, "The All-Particle Spectrum of Primary Cosmic Rays in the Wide Energy Range from 10^{14} to 10^{17} eV Observed with the Tibet-III Air-Shower Array," *Astrophysical Journal* 678, no. 2 (May 10, 2008): 1165–1179, doi:10.1086/529514.
16. A. D. Erlykin, S. K. Machavariani, and A. W. Wolfendale, "Puzzles of the Cosmic Ray Anisotropy," *Advances in Space Research* 63, no. 1 (January 1, 2019): 794–799, doi:10.1016/j.

asr.2018.09.042; Erlykin and Wolfendale, "Further Evidence Favouring the Single Source Model"; A. D. Erlykin and A. W. Wolfendale, "High-Energy Cosmic Gamma Rays from the 'Single Source,'" *Journal of Physics G: Nuclear and Particle Physics* 29, no. 4 (April 2003): 709–718, doi:10.1088/0954-3899/29/4/308; Erlykin and Wolfendale, "A Single Source of Cosmic Rays"; Erlykin and Wolfendale, "Spectral Features and Masses."

17. A. D. Erlykin and A. W. Wolfendale, "Interpretation of Features in the Cosmic Ray Proton and Helium Energy Spectra in Terms of a Local Source," *Journal of Physics G: Nuclear and Particle Physics* 42, no. 12 (November 9, 2015): id. 125201, doi:10.1088/0954-3899/42/12/125201; A. D. Erlykin and A. W. Wolfendale, "Secondary Cosmic Ray Nuclei in the Light of the Single Source Model and Comparison with Recent AMS-02 Data," *Journal of Physics G: Nuclear and Particle Physics* 43, no. 10 (August 23, 2016): id. 105201, doi:10.1088/0954-3899/43/10/105201.
18. Erlykin and Wolfendale, "Spectral Features and Masses."
19. A. D. Erlykin and A. W. Wolfendale, "High-Energy Cosmic Ray Spectroscopy. I. Status and Prospects," *Astroparticle Physics* 7, nos. 1–2 (June 1997): 1–13, doi:10.1016/S0927-6505(97)00014-5.
20. Craig, *In Quest of the Historical Adam*, 121. The summary to which Craig refers is in Hugh Ross, *Navigating Genesis: A Scientist's Journey Through Genesis 1–11* (RTB Press, 2014), 155–172.
21. Craig, *In Quest of the Historical Adam*, 124.
22. Craig, 124.
23. Craig, 124–125.
24. Craig, 125.
25. United States Census Bureau, *U.S. and World Population Clock*, accessed May 1, 2024, census.gov/popclock/.
26. Craig, *In Quest of the Historical Adam*, 128.
27. Craig, 129.
28. Craig, 104–131.
29. Joseph Miller, *Set Apart: How Genesis Stands Alone* (RTB Press, 2024).

Chapter 16: Life Before and After the Flood

1. Camilo Mora et al., "How Many Species Are There on Earth and in the Ocean?," *PLoS Biology* 9, no. 8 (August 23, 2011): e1001127, doi:10.1371/journal.pbio.1001127.
2. Mora et al., "How Many Species."
3. John C. Whitcomb, Jr. and Henry M. Morris, *The Genesis Flood: The Biblical Record and Its Scientific Implications* (Presbyterian & Reformed Publishing, 1961), 66–69.
4. John Woodmorappe, "The Biota and Logistics of Noah's Ark," *Proceedings of the International Conference on Creationism* 3, no. 55 (Creation Science Fellowship, 1994): 623–632, accessed November 21, 2023, digitalcommons.cedarville.edu/cgi/viewcontent.cgi?article=1366&context=icc_proceedings.
5. Ole Martin Moen and Katrien Devolder, "Palliative Farming," *Journal of Ethics* 26, no. 4 (September 13, 2022): 543–561, doi:10.1007/s10892-022-09404-7.
6. Nathaniel T. Jeanson, "Which Animals Were on the Ark with Noah?," *The Origin of Species After the Flood* (May 28, 2016), accessed November 21, 2023, answersingenesis.org/creation-science/baraminology/which-animals-were-on-the-ark-with-noah/.
7. Nathaniel T. Jeanson, "Getting Enough Genetic Diversity: How Species Arose After the Ark," (July 23, 2016), accessed December 28, 2024, answersingenesis.org/creation-science/baraminology/getting-enough-genetic-diversity/.

8. Henry M. Morris, *Biblical Creationism: What Each Book of the Bible Teaches About Creation and the Flood* (Mater Books, 2000), 36; Whitcomb and Morris, *The Genesis Flood*, 66–69 (in particular, figure 4 on p. 67 shows, for example, zebras and horses evolving from a single horse-like *pair* on board Noah's ark).
9. Don Batten, "Ligers and Wholphins? What Next?," *Creation* 22, no. 3 (June 2000): 28–33, accessed January 2, 2025, creation.com/ligers-and-wholphins-what-next; Bodie Hodge and Georgia Purdom, "Zonkeys, Ligers, and Wolphins, Oh My!," *Answers in Depth* 3 (2008): 71–73, accessed November 22, 2023, answersingenesis.org/hybrid-animals/zonkeys-ligers-and-wolphins-oh-my/.
10. Ming-Shan Wang et al., "A Polar Bear Paleogenome Reveals Extensive Ancient Gene Flow from Polar Bears into Brown Bears," *Nature Ecology and Evolution* 6 (July, 2022): 936–944, doi:10.1038/s41559-022-01753-8.
11. R. Joel Duff, Thomas R. Beatman, and David S. MacMillan III, "Dissent with Modification: How Postcreationism's Claim of Hyperrapid Speciation Opposes Yet Embraces Evolutionary Theory," *Evolution: Education and Outreach* 13, no. 9 (May 19, 2020), doi:10.1186/s12052-020-00124-w, accessed November 21, 2023, evolution-outreach.biomedcentral.com/articles/10.1186/s12052-020-00124-w.
12. Philip Kitcher, "Born-Again Creationism," in *Intelligent Design Creationism and Its Critics: Philosophical, Theological, and Scientific Perspectives*, ed. Robert T. Pennock (MIT Press, 2001), 259.
13. R. Laird Harris, Gleason L. Archer Jr., and Bruce K. Waltke, *Theological Wordbook of the Old Testament*, vol. 1 (Moody Press, 1980), 503–504.
14. Grant R. Lambert, "Was the Pre-Flood Animal Kingdom Vegetarian?," *Creation Research Society Quarterly* 20, no. 2 (1983): 88, creationresearch.org/crsq-1983-volume-20-number-2_was-the-preflood-animal-kingdom-vegetarian.
15. Erika Hagelberg, "Recombination or Mutation Rate Heterogeneity? Implications for Mitochondrial Eve," *Trends in Genetics* 19, no. 2 (February 2003): 84–90 (quote is on p. 84), doi:10.1016/S0168-9525(02)00048-3.
16. Phillip Endicott et al., "Evaluating the Mitochondrial Timescale of Human Evolution," *Trends in Ecology and Evolution* 24, no. 9 (September 2009): 515–521, doi:10.1016/j.tree.2009.04.006.
17. Qiaomei Fu et al., "A Revised Timescale for Human Evolution Based on Ancient Mitochondrial Genomes," *Current Biology* 23, no. 7 (April 8, 2013): 553–559, doi:10.1016/j.cub.2013.02.044.
18. Adrien Rieux et al., "Improved Calibration of the Human Mitochondrial Clock Using Ancient Genomes," *Molecular Biology and Evolution* 31, no. 10 (October 2014): 2780–2792, doi:10.1093/molbev/msu222.
19. G. David Poznik et al., "Sequencing Y Chromosomes Resolves Discrepancy in Time to Common Ancestor of Males Versus Females," *Science* 341, no. 6145 (August 2, 2013): 562–565, doi:10.1126/science.1237619.
20. Wei Wei et al., "A Calibrated Human Y-Chromosomal Phylogeny Based on Resequencing," *Genome Research* 23, no. 2 (February 2013): 388–395, doi:10.1101/gr.143198.112.
21. Fulvio Cruciani et al., "A Revised Root for the Human Y Chromosomal Phylogenetic Tree: The Origin of Patrilineal Diversity in Africa," *American Journal of Human Genetics* 88, no. 6 (June 10, 2011): 814–818, doi:10.1016/j.ajhg.2011.05.002.
22. Poznik et al., "Sequencing Y Chromosomes."
23. Rosaria Scozzari et al., "An Unbiased Resource of Novel SNP Markers Provides a New Chronology for the Human Y Chromosome and Reveals a Deep Phylogenetic Structure in Africa," *Genome Research* 24, no. 3 (March 2014): 535–544, doi:10.1101/gr.160788.133.

24. G. David Poznik et al., “Punctuated Bursts in Human Male Demography Inferred from 1,244 Worldwide Y-Chromosome Sequences,” *Nature Genetics* 48, no. 6 (June 2016): 593–599, doi:10.1038/ng.3559.
25. Iain McDonald, “Improved Models of Coalescence Ages of Y-DNA Haplogroups,” *Genes (Basel)* 12, no. 6 (June 4, 2021): id. 862, 21, doi:10.3390/genes12060862.
26. McDonald, “Improved Models of Coalescence Ages,” 21.
27. Eleanor F. Miller, Andrea Manica, and William Amos, “Global Demographic History of Human Populations Inferred from Whole Mitochondrial Genomes,” *Royal Society Open Science* 5, no. 8 (August 22, 2018): id. 180543, doi:10.1098/rsos.180543.
28. Sanni Översti and Jukka U. Palo, “Variation in the Substitution Rates Among the Human Mitochondrial Haplogroup U Sublineages,” *Genome Biology and Evolution* 14, no. 7 (July 2022): id. evac097, 6, doi:10.1093/gbe/evac097.
29. Krzysztof A. Cyran and Marek Kimmel, “Alternatives to the Wright-Fisher Model: The Robustness of Mitochondrial Eve Dating,” *Theoretical Population Biology* 78, no. 3 (November 2010): 165–172, doi:101016/kj.tpb.2010.06.001.
30. Översti and Palo, “Variation in the Substitution Rates,” 6.
31. Brenna M. Henn et al., “Characterizing the Time Dependency of Human Mitochondrial DNA Mutation Rate Estimates,” *Molecular Biology and Evolution* 26, no. 3 (March 2009): 713, doi:10.1093/molbev/msn244.
32. A. D. Erlykin and A. W. Wolfendale, “A Single Source of Cosmic Rays in the Range 10^{15}–10^{16}eV,” *Journal of Physics G: Nuclear and Particle Physics* 23, no. 8 (August 1997): 979, doi:10.1088/0954-3899/23/8/012; A. D. Erlykin and A. W. Wolfendale, “Recent Progress in EAS Measurements and the Single SNR Model of the Knee,” *Nuclear Physics B—Proceedings Supplements* 75, nos. 1–2 (March 1999): 305–308, doi:10.1016/S0920-5632(99)00274-1; A. D. Erlykin and A. W. Wolfendale, “Spectral Features and Masses in the PeV Region,” *Nuclear Physics B—Proceedings Supplements* 122 (July 2003): 209–212, doi:10.1016/S0920-5632(03)80380-8; A. D. Erlykin, S. K. Machavariani, and A. W. Wolfendale, “Puzzles of the Cosmic Ray Anisotropy,” *Advances in Space Research* 63, no. 1 (January 2019): 794–799, doi:10.1016/j.asr.2018.09.042.
33. S. E. Thorsett, et al., “Pulsar PSR B0656+14, The Monogem Ring, and the Origin of the ‘Knee’ in the Primary Cosmic Ray Spectrum,” *Astrophysical Journal* 592, no. 2 (August 1, 2003): L71–L73, doi:10.1086/377682.
34. Jonathan R. Knies, Manami Sasaki, and Paul P Plucinsky, “Suzaku Observations of the Monogem Ring and the Origin of Gemini Ha Ring,” *Monthly Notices of the Royal Astronomical Society* 477, no. 4 (July 2018): 4414–4422, doi:10.1093/mnras/sty915.
35. A. D. Erlykin and A. W. Wolfendale, “Further Evidence Favouring the Single Source Model for Cosmic Rays,” *Astroparticle Physics* 23, no. 1 (February 2005): 1–9, doi:10.1016/j.astropartphys.2004.09.009; E. G. Berezhko and L. T. Ksenofontov, “Composition of Cosmic Rays Accelerated in Supernova Remnants,” *Journal of Experimental and Theoretical Physics* 89, no. 3 (September 1999): 391–403, doi:10.1134/1.558996.
36. Renaud Kaeuffer et al., “Unexpected Heterozygosity in an Island Mouflon Population Founded by a Single Pair of Individuals,” *Proceedings of the Royal Society B: Biological Sciences* 274, no. 1609 (February 22, 2007): 527–533, doi:10.1098/rspb.2006.3743.
37. S. Elizabeth Alter, Seth D. Newsome, and Stephen R. Palumbi, “Pre-Whaling Genetic Diversity and Population Ecology in Eastern Pacific Gray Whales: Insights from Ancient DNA and Stable Isotopes,” *PLoS ONE* 7, no. 5 (May 9, 2012): id. e35039, doi:10.1371/journal.pone.0035039.
38. Juan P. Torres-Florez et al., “High Genetic Diversity in a Small Population: The Case of

Chilean Blue Whales," *Ecology and Evolution* 4, no. 8 (April 2014): 1398–1412, doi:10.1002/ece3.998.

39. Hiroki Goto et al., "A Massively Parallel Sequencing Approach Uncovers Ancient Origins and High Genetic Variability of Endangered Przewalski's Horses," *Genome Biology and Evolution* 3 (July 29, 2011): 1096–1106, doi:10.1093/gbe/evr067.
40. Catherine Lippé, Pierre Dumont, and Louis Bernatchez, "High Genetic Diversity and No Inbreeding in the Endangered Copper Redhorse, *Moxostoma hubbsi* (Catostomidae, Pisces): The Positive Sides of a Long Generation Time," *Molecular Ecology* 15, no. 7 (June 2006): 1769–1780, doi:10.1111/j.1365-294X.2006.02902.x.
41. Steffen Roth and Robert Jehle, "High Genetic Diversity of Common Toad (*Bufo bufo*) Populations Under Strong Natural Fragmentation on a Northern Archipelago," *Ecology and Evolution* 6, no. 6 (March 2016): 1626–1636, doi:10.1002/ece3.1957.
42. Jaana Kekkonen, Mikael Wikström, and Jon E. Brommer, "Heterozygosity in an Isolated Population of a Large Mammal Founded by Four Individuals Is Predicted by an Individual-Based Genetic Model," *PLoS ONE* 7, no. 9 (September 20, 2012): id. e43482, doi:10.1371/journal.pone.0043482; Suzanne L. Nelson, Scott A. Taylor, and Jon D. Reuter, "An Isolated White-Tailed Deer (*Odocoileus virginianus*) Population on St. John, US Virgin Islands Shows Low Inbreeding and Comparable Heterozygosity to Other Larger Populations," *Ecology and Evolution* 11, no. 6 (March 2021): 2775–2781, doi:10.1002/ece3.7230.
43. Frank Hailer et al., "Bottlenecked but Long-Lived: High Genetic Diversity Retained in White-Tailed Eagles upon Recovery from Population Decline," *Biology Letters* 2, no. 2 (June 22, 2006): 316–319, doi:10.1098/rsbl.2006.0453.
44. Teresa L. Santos et al., "Conservation Genetic Assessment of Savannah Elephants (*Loxodonta africana*) in the Greater Kruger Biosphere, South Africa," *Genes (Basel)* 10, no. 10 (October 5, 2019): id. 779, doi:10.3390/genes10100779; Torres-Florez et al., "High Genetic Diversity," 1398–1412; Lippé, Dumont, and Bernatchez, "High Genetic Diversity and No Inbreeding," 1769–1780.
45. Richard J. Wang et al., "Human Generation Times Across the Past 250,000 Years," *Science Advances* 9, no. 1 (January 6, 2023): id. eabm7047, doi:10.1126/sciadv.abm7047.

Chapter 17: The Source of Earth's Biodeposits

1. Walter T. Brown, *In the Beginning: Compelling Evidence for Creation and the Flood*, 5th ed. (Center for Scientific Creation, 1989), 58–84; Jonathan Sarfati, *Refuting Compromise: A Biblical and Scientific Refutation of "Progressive Creationism" (Billions of Years), as Popularized by Astronomer Hugh Ross*, 2nd ed. (Creation Book Publishers, 2011), 273–274, 279–280, 368–370; Ken Ham and Bodie Hodge, *A Flood of Evidence: 40 Reasons Noah and the Ark Still Matter* (Master Books, 2016), 117–122.
2. Central Energy Resources Science Center, U. S. Geological Survey, *World Oil and Gas Resource Assessments* (April 26, 2019), accessed December 9, 2023, usgs.gov/centers/central-energy-resources-science-center/science/world-oil-and-gas-resource-assessments; Amy Myers Jaffe, Kenneth B. Medlock III, and Ronald Soligo, *The Status of World Oil Reserves: Conventional and Unconventional Resources in the Future Supply Mix*, Baker Institute for Public Policy, Rice University (October 2011): 11–26, accessed January 2, 2025. The full paper can be downloaded for free at bitly/3YxaV82; Nick A. Owen, Oliver R. Inderwildi, and David A. King, "The Status of Conventional World Oil Reserves—Hype or Cause for Concern?," *Energy Policy* 38, no. 8 (August 2010): 4743–4749, doi:10.1016/j.enpol.2010.02.026; Richard Nehring, "Traversing the Mountaintop: World Fossil Fuel

Production to 2050," *Philosophical Transactions of the Royal Society of London B, Biological Sciences* 364, no. 1532 (October 27, 2009): 3067–3079, doi:10.1098/rstb.2009.0170; Thomas S. Ahlbrandt and P. J. McCabe, "Global Petroleum Resources: A View to the Future," *Geotimes* 47, no. 11 (November 2002): 14–18; H-H. Rogner, "An Assessment of World Hydrocarbon Resources," *Annual Review of Energy and the Environment* 22 (November 1997): 217–262, doi:10.1146/annurev.energy.22.1.217.

3. Qais M. Abdulhamid et al., "Thermal Cracking of Al-Dora Asphalt for the Simultaneous Production of Light Fuel and Activated Carbon for Desulfurization Process," *Journal of Analytical and Applied Pyrolysis* 173 (August 2023): id. 106072, doi:10.1016/j.jaap.2023.106072; Ray Boswell et al., "6–Natural Gas Hydrates: Status of Potential as an Energy Source," in *Future Energy: Improved, Sustainable and Clean Options for our Planet*, 3rd ed. (2020): 111–131, doi:10.1016/B978-0-08-102886-5.00006-2; Central Energy Resources Science Center, U. S. Geological Survey, *World Oil and Gas Resource Assessments* (April 26, 2019), accessed December 9, 2023, usgs.gov/centers/central-energy-resources-science-center/science/world-oil-and-gas-resource-assessments; Energy Information Administration, "Future Supply Potential of Natural Gas Hydrates," in *Natural Gas 1998: Issues and Trends* (EIA, 1998): 74, eia.gov/pub/oil_gas/natural_gas/analysis_publications/natural_gas_1998_issues_trends/pdf/chapter3.pdf.
4. Kurt Bucher and Rodney Grapes, "Metamorphic Rocks," in *Petrogenesis of Metamorphic Rocks*, 8th ed. (Springer, 2011), 24, doi:10.1007/978-3-540-74169-5_2, books.google.com/books?id=FFcHarai9GEC&pg=PA21#v=onepage&q&f=false.
5. Harvey Blatt and Robert J. Tracy, *Petrology: Igneous, Sedimentary, and Metamorphic*, 2nd ed. (W. H. Freeman and Co., 1995), 295–300.
6. Bradley R. Hacker, Peter B. Kelemen, and Mark D. Behn, "Continental Lower Crust," *Annual Review of Earth and Planetary Sciences* 43 (May 2015): 167–205, doi:10.1146/annurev-earth-050212-124117.
7. W. M. White and E. M. Klein, "4.13–Composition of the Oceanic Crust," *Treatise on Geochemistry*, 2nd ed., vol. 4 (2014): 457–496, doi:10.1016/B978-0-08-095975-7.00315-6.
8. C. Michael Hogan et al., "Limestone," *Encyclopedia of Earth*, ed. Cutler J. Cleveland (Environmental Information Coalition, National Council for Science and the Environment, October 16, 2011), accessed December 13, 2023, editors.eol.org/eoearth/wiki/Limestone; Bucher and Grapes, *Petrogenesis of Metamorphic Rocks*, 24.
9. Kazuhisa Miyamoto, ed., "Energy Conversion by Photosynthetic Organisms," in *Renewable Biological Systems for Alternative Sustainable Energy Production*, FAO Agricultural Services Bulletin, 128 (FAO Corporate Document Repository, 1997): chap. 2, section 2.1.1, accessed December 14, 2023, fao.org/docrep/w7241e/w7241e06.htm#TopOfPage.
10. Mathieu Dellinger et al., "High Rates of Rock Organic Carbon Oxidation Sustained as Andean Sediment Transits the Amazon Foreland-Floodplain," *Proceedings of the National Academy of Sciences, USA* 120, no. 39 (September 19, 2023): id. e2306343120, doi:10.1073/pnas.2306343120; Steven T. Petsch, "12.8–Weathering of Organic Carbon," *Treatise on Geochemistry*, 2nd ed. (2014), 217–238, chap. 12, doi:10.1016/B978-0-08-095975-7.01013-5; S. T. Petsch, K. J. Edwards, and T. I. Eglinton, "Microbial Transformations of Organic Matter in Black Shales and Implications for Global Biogeochemical Cycles," *Palaeogeography, Palaeoclimatology, Palaeoecology* 219, nos. 1–2 (April 11, 2005): 157–170, doi:10.1016/j.palaeo.2004.10.019; S. T. Petsch, T. I. Eglinton, and K. J. Edwards, "^{14}C-Dead Living Biomass: Evidence for Microbial Assimilation of Ancient Organic Carbon During Shale Weathering," *Science* 292, no. 5519 (May 11, 2001): 1127–1131, doi:10.1126/science.1058332.

11. Dellinger et al., "High Rates of Rock Organic."
12. Brian Groombridge and Martin D. Jenkins, *World Atlas of Biodiversity: Earth's Living Resources in the 21st Century* (University of California Press, 2002), 10–11.
13. See, for example, Brown, *In the Beginning*, 58–84; Sarfati, *Refuting Compromise*, 273–274, 279–280, 368–370; Ham and Hodge, *A Flood of Evidence*, 117–122.
14. Takayuki Tashiro et al., "Early Trace of Life from 3.95 Ga Sedimentary Rocks in Labrador, Canada," *Nature* 549, no. 7673 (September 28, 2017): 516–518, doi:10.1038/nature24019; Minik T. Rosing, "13C-Depleted Carbon Microparticles in >3700-Ma Sea-Floor Sedimentary Rocks from West Greenland," *Science* 283, no. 5402 (January 29, 1999): 674–676, doi:10.1126/science.283.5402.674; S. Flögel et al., "Simulating the Biogeochemical Effects of Volcanic CO_2 Degassing on the Oxygen-State of the Deep Ocean During the Cenomanian/Turonian Anoxic Event (OAE2)," *Earth and Planetary Science Letters* 305, nos. 3–4 (May 15, 2011): 371–374, doi:10.1016/j.epsl.2011.03.018; David B. Finkelstein, Lisa M. Pratt, and Simon C. Brassell, "Can Biomass Burning Produce a Globally Significant Carbon-Isotope Excursion in the Sedimentary Record?," *Earth and Planetary Science Letters* 250, nos. 3–4 (October 30, 2006): 501–510, doi:10.1016/j.espl.2006.08.010; Denis Lacelle et al., "Geomicrobiology and Occluded O_2–CO_2–Ar Gas Analyses Provide Evidence of Microbial Respiration in Ancient Terrestrial Ground Ice," *Earth and Planetary Science Letters* 306, nos. 1–2 (June 1, 2011): 46–54, doi:10.1016/j.epsl.2011.03.023; Jennifer L. Macalady et al., "Dominant Microbial Populations in Limestone-Corroding Stream Biofilms, Frasassi Cave System, Italy," *Applied and Environmental Microbiology* 72, no. 8 (August 2006): 5596–5609, doi:10.1128/AEM.00715-06; K. M. Meyer et al., "$d^{13}C$ Evidence That High Primary Productivity Delayed Recovery from End-Permian Mass Extinction," *Earth and Planetary Science Letters* 302, nos. 3–4 (February 1, 2011): 378–384, doi:10.1016/j.epsl.2010.12.033.
15. Hogan, "Limestone," *Encyclopedia of Earth.*
16. Blatt and Tracy, *Petrology*, 295–300.
17. David Wright, "Timeline for the Flood," *Feedback, Answers in Genesis*, March 9, 2012, accessed December 15, 2023, answersingenesis.org/bible-timeline/timeline-for-the-flood/; Brian Thomas, "When Did Noah's Flood Happen?," *Acts & Facts* 47, no. 3, Institute for Creation Research (February 28, 2018), icr.org/article/when-did-noahs-flood-happen/; John Osgood, "The Date of Noah's Flood," *Creation* 4, no. 1 (March 1981): 10–13, creation.com/the-date-of-noahs-flood.
18. Michael Oard, "The Ice Age and the Genesis Flood," *Acts & Facts: Impact* 16, no. 6 (June 1, 1987), accessed May 3, 2024, icr.org/article/ice-age-genesis-flood; John D. Morris, "The Ice Age: Causes and Consequences," *Acts & Facts: Back to Genesis* 38, no. 8 (August 1, 2009): 15, accessed May 3, 2024, icr.org/article/ice-age-causes-consequences; Andrew A. Snelling and Mike Matthews, "When Was the Ice Age in Biblical History?," *Answers Magazine*, April 1, 2013, 46–52, accessed May 3, 2024, answersingenesis.org/environmental-science/ice-age/when-was-the-ice-age-in-biblical-history/; Larry Vardiman, "An Analytical Young-Earth Flow Model of the Ice Sheet Formation During the 'Ice-Age,'" *Proceedings of the International Conference on Creationism* 3, no. 60 (1994): 561–568, accessed May 3, 2024, digitalcommons.cedarville.edu/icc_proceedings/vol3/iss1/60; Larry Vardiman, "What Started the Ice Age?," *Answers Magazine*, April 1, 2013, 53–55, accessed May 3, 2024, answersingenesis.org/environmental-science/ice-age/what-started-the-ice-age/.
19. P. Fretwell et al., "Bedmap2: Improved Ice Bed, Surface and Thickness Datasets for Antarctica," *The Cryosphere* 7, no. 1 (February 28, 2013): 375–393, doi:10.5194/tc-7-375-2013.
20. Cyril Palerme et al., "Evaluation of Current and Projected Antarctic Precipitation in CMIP5

Models," *Climate Dynamics* 48, nos. 1–2 (January 2017): 225–239, doi:10.1007/s00382-016-3071-1; Cyril Palerme et al., "How Much Snow Falls on the Antarctic Ice Sheet?," *The Cryosphere* 8, no. 4 (August 2014): 1577–1587, doi:10.5194/tc-8-1577-2014.

21. F. Lemonnier et al., "CloudSat-Inferred Vertical Structure of Snowfall Over the Antarctic Continent," *Journal of Geophysical Research: Atmospheres* 125, no. 2 (January 27, 2020): id. e2019JD031399, doi:10.1029/2019JD031399.

Chapter 18: Rainbows, Physics, and Noah's Flood

1. John C. Dillow, *The Waters Above: Earth's Pre-Flood Vapor Canopy*, rev. ed. (Moody Press, 1981); Henry M. Morris, *The Genesis Record: A Scientific and Devotional Commentary on the Book of Beginnings* (Baker Books, 1976), 60, 191; John C. Whitcomb and Henry M. Morris, *The Genesis Flood: The Biblical Record and Its Scientific Implications* (Presbyterian & Reformed Publishing, 1982), 239–244; NeverThirsty: Like the Master Ministries, "Was There Rain Before and After Noah's Flood in Genesis?," accessed January 2, 2025, neverthirsty.org/bible-qa/qa-archives/question/was-there-rain-before-and-after-noahs-flood-in-genesis/.
2. R. Laird Harris, Gleason L. Archer Jr., and Bruce K. Waltke, *Theological Wordbook of the Old Testament*, vol. 1 (Moody Press, 1980), 500–501.
3. Harris, Archer, and Waltke, *Theological Wordbook*, 17.
4. Sanjoy M. Som et al., "Earth's Air Pressure 2.7 Billion Years Ago Constrained to Less Than Half of Modern Levels," *Nature Geoscience* 9, no. 6 (June 2016): 448–451, doi:10.1038/ngeo2713; Sanjoy M. Som et al., "Air Density 2.7 Billion Years Ago Limited to Less Than Twice Modern Levels by Fossil Raindrop Imprints," *Nature* 484, no. 7394 (April 19, 2012): 359–362, doi:10.1038/nature10890; J. K. Zimmerman and S. M. Som, "Refining Raindrop Paleobarometry: A Status Report," *American Geophysical Union*, Fall Meeting 2016, abstract #PP31A-2259 (December 2016), ui.adsabs.harvard.edu/abs/2016AGUFMPP31A2259Z/abstract; C. Glotzbach and C. Brandes, "Air Density of the Permian Atmosphere: Constraints from Lithified Raindrop Imprints," *Palaeogeography, Palaeoclimatology, Palaeoecology* 409 (September 1, 2014): 280–289, doi:10.1016/j.palaeo.2014.05.024.
5. Som et al., "Earth's Air Pressure 2.7 Billion Years Ago."
6. Jeremiah 33:15, NASB.
7. B. P. Abbott et al., LIGO Scientific Collaboration and Virgo Collaboration, "GW170817: Observation of Gravitational Waves from a Binary Neutron Star Inspiral," *Physical Review Letters* 119, no. 16 (October 20, 2017): id. 161101, doi:10.1103/PhysRevLett.119.161101; B. P. Abbott et al., LIGO Scientific Collaboration and Virgo Collaboration, Fermi Gamma-Ray Burst Monitor, and INTEGRAL, "Gravitational Waves and Gamma-Rays from a Binary Neutron Star Merger: GW170817 and GRB 170817A," *Astrophysical Journal Letters* 848, no. 2 (October 20, 2017): id. L13, doi:10.3847/2041-8213/aa920c; Jun-Jie Wei et al., "Limits on the Neutrino Velocity, Lorentz Invariance, and the Weak Equivalence Principle with TeV Neutrinos from Gamma-Ray Bursts," *Journal of Cosmology and Astroparticle Physics* 2016, no. 8 (August 2016): id. 031, doi:10.1088/1475-7516/2016/08/031.
8. H. M. Altschuler, "Commission 1. Radio Measurement Methods and Standards," *Radio Science* 8, no. 6 (June 1973): 567–572, doi:10.1029/RS008i006p00567.
9. I provide a detailed explanation for why such a tiny change in the Sun's luminosity would be catastrophic for life in *Improbable Planet: How Earth Became Humanity's Home* (Baker Books, 2016), 143–164.
10. George F. R. Ellis, "The Anthropic Principle: Laws and Environments," in *The Anthropic Principle*, ed. F. Bertola and U. Curi (Cambridge University Press, 1993), 30; Yun Liu and

Ji Miao, "An Emerging Role of Defective Copper Metabolism in Heart Disease," *Nutrients* 14, no. 3 (February 7, 2022): id. 700, doi:10.3390/nu14030700; H. R. Marston, Shirley H. Allen, and S. L. Swaby, "Iron Metabolism in Copper-Deficient Rats," *British Journal of Nutrition* 25, no. 1 (January 1971): 15–30, doi:10.1079/BJN19710062; K. W. J. Wahle and N. T. Davies, "Effect of Dietary Copper Deficiency in the Rat on Fatty Acid Composition of Adipose Tissue and Desaturase Activity of Liver Microsomes," *British Journal of Nutrition* 34, no. 1 (July 1975): 105–112, doi:10.1017/s000711457500013x; Walter Mertz, "The Newer Essential Trace Elements, Chromium, Tin, Vanadium, Nickel, and Silicon," *Proceedings of the Nutrition Society* 33 (December 1974): 307–313, doi:10.1079/pns19740054.

11. Michael T. Murphy et al., "Fundamental Physics with ESPRESSO: Precise Limit on Variations in the Fine-Structure Constant Towards the Bright Quasar HE 0515-4414," *Astronomy & Astrophysics* 658 (February 2022): id. A123, doi:10.1051/0004-6361/202142257; T. D. Le, "Stringent Limit on Space-Time Variation of Fine-Structure Constant Using High-Resolution of Quasar Spectra," *Heliyon* 6, no. 9 (September 23, 2020): id. e05011, doi:10.1016/j.heliyon.2020.e05011; Haoran Liang and Zhe Wu, "Measuring the Fine-Structure Constant on Quasar Spectra: High Spectral Resolution Gains More Than Large Size of Moderate Spectral Resolution Spectra," *Proceedings of the SPIE 12644, International Workshop on Frontiers of Graphics and Image Processing*, FGIP 2022 (May 3, 2023): id. 1264408, doi:10.1117/12.2670012; Michael T. Murphy, Adrian L. Malec, and J. Xavier Prochaska, "Precise Limits on Cosmological Variability of the Fine-Structure Constant with Zinc and Chromium Quasar Absorption Lines," *Monthly Notices of the Royal Astronomical Society* 461, no. 3 (September 21, 2016): 2461–2479, doi:10.1093/mnras/stw1482; S. A. Levshakov et al., "An Upper Limit to the Variation in the Fundamental Constants at Redshift $z = 5.2$," *Astronomy & Astrophysics: Letters* 540 (April 2012): id. L9, doi:10.1051/0004-6361/201219042; Franco D. Albareti et al., "Constraint on the Time Variation of the Fine-Structure Constant with the SDSS-III/BOSS DR12 Quasar Sample," *Monthly Notices of the Royal Astronomical Society* 452, no. 4 (October 1, 2015): 4153–4168, doi:10.1093/mnras/atv1406; S. M. Kotuš, M. T. Murphy, and R. F. Carswell, "High-Precision Limit on Variation in the Fine-Structure Constant from a Single Quasar Absorption System," *Monthly Notices of the Royal Astronomical Society* 464, no. 3 (January 2017): 3679–3703, doi:10.1093/mnras/stw2543; R. Srianand et al., "Limits on the Time Variation of the Electromagnetic Fine-Structure Constant in the Low Energy Limit from Absorption Lines in the Spectra of Distant Quasars," *Physical Review Letters* 92, no. 12 (March 26, 2004): id. 121302, doi:10.1103/PhysRevLett.92.121302.
12. T. Rosenband et al., "Frequency Ratio of Al+ and Hg+ Single-Ion Optical Clocks; Metrology at the 17th Decimal Place," *Science* 319, no. 5871 (March 28, 2008): 1808–1812, doi:10.1126/science.1154622.
13. Ira Wasserman and Kenneth Brecher, "Time Invariance of Planck's Constant," *Physical Review Letters* 41, no. 14 (October 2, 1978): 920–923, doi:10.1103/PhysRevLett.41.920.
14. S. John Pirt, "The Thermodynamic Efficiency (Quantum Demand) and Dynamics of Photosynthetic Growth," *New Phytologist* 102, no. 1 (January 1986): 3–37, doi:10.1111/j.1469-8137.1986.tb00794.x; Chao-Ping Hsu, "The Electronic Couplings in Electron Transfer and Excitation Energy Transfer," *Accounts of Chemical Research* 42, no. 4 (April 21, 2009): 509–518, doi:10.1021/ar800153f.
15. Gravity works to collapse a star. Electromagnetic radiation works to expand it. A change as small as 1 part in 10^{40} (one part in ten thousand trillion trillion trillion) in the ratio of the gravitational to electromagnetic coupling constants will destabilize a star.
16. Yu V. Petrov et al., "Natural Nuclear Reactor at Oklo and Variation of Fundamental

Constants: Computation and Neutronics of a Fresh Core," *Physical Review C* 74, no. 6 (December 14, 2006): id. 064610, doi:10.1103/PhysRevC.74.064610; V. V. Flambaum and E. V. Shuryak, "Limits on Cosmological Variation of Strong Interaction and Quark Masses from Big Bang Nucleosynthesis, Cosmic, Laboratory, and Oklo Data," *Physical Review D* 65, no. 10 (May 2002): id. 103503, doi:10.1103/PhysRevD.65.103503; V. F. Dmitriev and V. V. Flambaum, "Limits on Cosmological Variation of Quark Masses and Strong Interaction," *Physical Review D* 67, no. 6 (March 26, 2003): id. 063513, doi:10.1103/PhysRevD.67.063513; Yasunori Fujii et al., "The Nuclear Interaction at Oklo 2 Billion Years Ago," *Nuclear Physics B* 573, nos. 1–2 (May 1, 2000): 377–401, doi:10.1016/S0550-3213(00)00038-9.

17. J. Kobayashi, A. Ogino, and S. Inouye, "Measurement of the Variation of Electron-to-Proton Mass Ratio Using Ultracold Molecules Produced from Laser-Cooled Atoms," *Nature Communications* 10 (August 21, 2019): id. 3771, doi:10.1038/s41467-019-11761-1.
18. N. Huntemann et al., "Improved Limit on a Temporal Variation of m_p/m_e from Comparisons of Yb^+ and Cs Atomic Clocks," *Physical Review Letters* 113, no. 21 (November 17, 2014): id. 210802, doi:10.1103/PhysRevLett.113.210802.
19. R. M. Godun et al., "Frequency Ratio of Two Optical Clock Transitions in $^{171}Yb^+$ and Constraints on the Time Variation of Fundamental Constants," *Physical Review Letters* 113, no. 21 (November 21, 2014): id. 210801, doi:10.1103/PhysRevLett.113.210801.
20. Liliane Biskupek, Jürgen Müller, and Jean-Marie Torre, "Benefit of New High-Precision LLR Data for the Determination of Relativistic Parameters," *Universe* 7, no. 2 (February 3, 2021): id. 34, doi:10.3390/universe7020034.
21. Biskupek, Müller, and Torre, "Benefits of New High-Precision LLR Data."
22. Julija Bagdonaite et al., "A Stringent Limit on a Drifting Proton-to-Electron Mass Ratio from Alcohol in the Early Universe," *Science* 339, no. 6115 (January 4, 2013): 46–48, doi:10.1126/science.1224898.
23. S. A. Levshakov et al., "An Upper Limit to the Variation in the Fundamental Constants at Redshift $z = 5.2$," *Astronomy & Astrophysics: Letters* 540 (April 2012): id. L9, doi:10.1051/0004-6361/201219042.
24. Earl Patrick Bellinger and Jørgen Christensen Dalsgaard, "Astroseismic Constraints on the Cosmic-Time Variation of the Gravitational Constant from an Ancient Main-Sequence Star," *Astrophysical Journal Letters* 887, no. 1 (December 3, 2019): id. L1, doi:10.3847/2041-8213/ab43e7.
25. W. W. Zhu et al., "Tests of Gravitational Symmetries with Pulsar Binary J1713+0747," *Monthly Notices of the Royal Astronomical Society* 482, no. 3 (January 2019): 3249–3260, doi:10.1093/mnras/sty2905.
26. Alfio Bonanno and Hans-Erich Fröhlich, "A New Helioseismic Constraint on a Cosmic-Time Variation of *G*," *Astrophysical Journal Letters* 893, no. 2 (April 21, 2020): id. L35, doi:10.3847/2041-8213/ab86b9.

Chapter 19: Purposes and Benefits of Noah's Flood

1. Numbers 13; Deuteronomy 1–3, 9; Joshua 12–15, 17–18; Judges 1; 1 Samuel 17, 21; 2 Samuel 21; 1 Chronicles 20.
2. Hugh Ross, *Navigating Genesis: A Scientist's Journey Through Genesis 1–11* (RTB Press, 2014), 131–138, 233–234.
3. Dhongil Lim et al., "Enhancement of Volcanic Eruption in Mid-Ocean Ridge During the Last Deglaciation: New Sedimentary Evidence in the Middle Part of Central Indian Ridge," *Marine Geology* 440 (October 2021): id. 106574, doi:10.1016/j.margeo.2021.106574;

Summer Praetorius et al., "Interaction Between Climate, Volcanism, and Isostatic Rebound in Southeast Alaska During the Last Deglaciation," *Earth and Planetary Science Letters* 452 (October 15, 2016): 79–89, doi:10.1016/j.epsl.2016.07.033; Peter M. Abbott and Siwan M. Davies, "Volcanism and the Greenland Ice-Cores: The Tephra Record," *Earth-Science Reviews* 115, no. 3 (November 2012): 173–191, doi:10.1016/j.earscirev.2012.09.001; Peter M. Abbott et al., "A Detailed Framework of Marine Isotope Stages 4 and 5 Volcanic Events Recorded in Two Greenland Ice-Cores," *Quaternary Science Reviews* 36 (March 12, 2012): 59–77, doi:10.1016/j.quascirev.2011.05.001; S. Kutterolf et al., "Milankovitch Frequencies in Tephra Records at Volcanic Arcs: The Relation of Kyr-Scale Cyclic Variations in Volcanism to Global Climate Changes," *Quaternary Science Reviews* 204 (January 15, 2019): 1–16, doi:10.1016/j.quascirev.2018.11.004; Julie C. Schindlbeck et al., "100-Kyr Cyclicity in Volcanic Ash Emplacement: Evidence from a 1.1 Myr Tephra Record from the NW Pacific," *Scientific Reports* 8 (March 23, 2018): id. 4440, doi:10.1038/s41598-018-22595-0.

4. Hugh Ross, *Rescuing Inerrancy: A Scientific Defense* (RTB Press, 2024).

Appendix: Dating Early Human History

1. Norman E. Holden, "Total Half-Lives for Selected Nuclides," *Pure and Applied Chemistry* 62, no. 5 (1990): 941–958, doi:10.1351/pac199062050941.
2. D. A. Green, "A Revised Catalogue of 294 Galactic Supernova Remnants," *Journal of Astrophysics and Astronomy* 40, no. 36 (August 23, 2019), doi:10.1007/s12036-019-9601-6; V. Borka, "Spectral Indices of Galactic Radio Loops Between 1420, 820, and 408 MHz," *Monthly Notices of the Royal Astronomical Society* 376, no. 2 (April 2007): 634–644, doi:10.1111/j.1365-2966.2007.11499.x; Vo Hong Minh Phan et al., "Stochasticity of Cosmic Rays from Supernova Remnants and the Ionization Rates in Molecular Clouds," *Physical Review D* 107, no. 12 (June 8, 2023): id. 123006, doi:10.1103/PhysRevD.107.123006; Siddhartha Gupta et al., "Realistic Modelling of Wind and Supernovae Shocks in Star Clusters: Addressing ^{22}Ne/^{20}Ne and Other Problems in Galactic Cosmic Rays," *Monthly Notices of the Royal Astronomical Society* 493, no. 3 (April 2020): 3159–3177, doi:10.1093/mnras/staa286; R. B. Firestone, "Observation of 23 Supernovae That Exploded <300 pc from Earth During the Past 300 kyr," *Astrophysical Journal* 789, no. 1 (June 11, 2014): id. 29, doi:10.1088/0004-637X/789/1/29; L. D. Anderson et al., "Supernova Remnant Candidates Discovered by the SARAO MeerKAT Galactic Plane Survey," *Astronomy & Astrophysics* 693 (January 2025): id. A247, doi:10.1051/0004-6361/202451038; Alexis Quintana, Nicholas J. Wright, and Juan Martinez-Garcia, "The Structure and Star Formation History of the Local Milky Way as Revealed by Its OB Stars," European Astronomical Society Annual Meeting, held July 1–5, 2024 in Padova, Italy. Session S4 : Gaia: The (TWO) Billion Star Galaxy Census: from the Solar System to our Milky Way and Beyond., ePoster, id. 623, ui.adsabs.harvard.edu/abs/2024eas..conf..623Q/abstract.
3. R. L. K. Fullagar, D. M. Price, and L. M. Head, "Early Human Occupation of Northern Australia: Archaeology and Thermoluminescence Dating of Jinmium Rock-Shelter, Northern Territory," *Antiquity* 70, no. 270 (December 1996): 751–773, doi:10.1017/S0003598X00084040.
4. Richard Roberts et al., "Optical and Radiocarbon Dating at Jinmium Rock Shelter in Northern Australia," *Nature* 393 (May 28, 1998): 358–362, doi:10.1038/30718.
5. Roberts et al., "Optical and Radiocarbon Dating."
6. Bruno David et al., "New Optical and Radiocarbon Dates from Ngarrabullgan Cave, a Pleistocene Archaeological Site in Australia: Implications for the Comparability of Time Clocks and for the Human Colonization of Australia," *Antiquity* 71, no. 271 (March 1997):

183–188, doi:10.1017/S0003598X00084672.

7. Firestone, "Observation of 23 Supernovae."
8. Anderson et al., "Supernova Remnant Candidates"; Quintana, Wright, and Martinez-Garcia, "The Structure and Star Formation History"; Firestone, "Observation of 23 Supernovae."
9. Anderson et al., "Supernova Remnant Candidates"; Quintana, Wright, and Martinez-Garcia, "The Structure and Star Formation History"; Firestone, "Observation of 23 Supernovae."

Index

About the Author

Hugh Ross is senior scholar and founder of Reasons to Believe (RTB), an organization dedicated to demonstrating the compatibility of science and the Christian faith.

With a degree in physics from the University of British Columbia and a PhD in astronomy from the University of Toronto, Hugh—initially a skeptic, always curious, and eventually a Christian—continued his research on quasars and galaxies as a postdoctoral fellow at the California Institute of Technology. After five years there, he transitioned to full-time ministry. In addition to founding and leading RTB, he remains on the pastoral staff at Christ Church Sierra Madre. His writings include journal and magazine articles, blogs, and numerous books—*The Creator and the Cosmos*, *Why the Universe Is the Way It Is*, and *Designed to the Core*, among others. He has spoken on hundreds of university campuses as well as at conferences and churches around the world.

Hugh lives in Southern California with his wife, Kathy.

About Reasons to Believe

Reasons to Believe (RTB) exists to reveal God in science. Based in Covina, California, RTB was established in 1986 and since then has taken scientific evidence for the God of the Bible across the US and around the world. Our ongoing work is providing content for all who desire to explore the connection between science and the Christian faith.

RTB is unique in its range of resources. The curious can explore articles, podcasts, and videos. Those who want to learn more can delve into books, in-person and live streamed events, and online courses. Donors enable us to continue this important work.

For more information, visit reasons.org.

For inquiries, contact us via:
818 S. Oak Park Rd.
Covina, CA 91724
(855) REASONS | (855) 732-7667
ministrycare@reasons.org

@rtb_official